Earth's Voyage Through Time

To Professor L. J. Wills, a pioneer in palaeogeology, with affection and respect.

Earth's

Voyage Through Time

David Dineley

BOOK CLUB ASSOCIATES
LONDON

This edition published 1974 by
Book Club Associates
By arrangement with Granada Publishing Limited

Copyright © 1973 by David Dineley

Printed in Great Britain by
Fletcher & Son Ltd, Norwich

CONTENTS

LIST OF ILLUSTRATIONS

LIST OF PLATES

PREFACE

MANY an author has welcomed the chance to write a preface after the typescript of his book has gone to the press. If a new discovery had been made since the text was written it could be mentioned in a rather stop-press way and the author could feel that he was as up to date as possible. For the writer of this book the preface offers no such opportunity and he has to admit that a great fund of new information about the state of our planet has appeared since the last word of the last chapter was written. The pace of geological discovery and of our reappraisal of the evolution of this small planet and its solar system is now so rapid that no author can cram all the latest scientific news into the page or two that precedes his text. While the hypotheses, ideas and evidence that geologists have been mulling over in the last few years are likely to remain reasonably intact for a little longer, they certainly will have to be modified in the next few years.

New theories and concepts sometimes appear from the most unexpected quarters or owe their origins to discoveries in seemingly obscure fields. Who would have thought that a vital hypothesis about the long evolution of the crust of the earth would develop from the exploration of the deep ocean floor?

There have been many 'Theories of the Earth' and accounts of the changing days since life began to evolve on this planet. No single theory seems to account satisfactorily for the behaviour of the earth in every respect. We can visualise several different models of how it may be in nature and one can take one's choice. There is simply not enough room on the pages of this book to describe and discuss all the alternatives. The story as construed from the models offered is, I hope, reasonable enough for the moment as the mere outline of a long and complicated plot, involving more characters and 'plays within a play' than most of us can recognise.

The writing of these pages has been made the more pleasant and the finished article made the better by the help and encouragement of my colleagues in the Department of Geology at the University of

Bristol. Mrs. A. Gregory and Mr. R. Godwin helped draft and prepare the text figures and Mrs. J. Rowland typed from my scarcely legible manuscript. To them all I extend my sincere thanks.

D.L.D.

CHAPTER I

Debate About the Earth

THERE cannot be many men on earth who have not wondered how and where it all began. Throughout history, and no doubt long before history itself commenced, men have been given to wondering about the nature of things, where they all came from and what was the purpose of the world. Few of us can remain unimpressed by the sky at night, the ferocity of a hurricane or the sinister expressions of power in a volcano. It all means something. It means different things to different people. Cavemen long ago would perhaps have seen the work of gods, spirits and demons in these splendid phenomena; the city-dweller may regard them as mere indications that the world, the solar system and the universe consist of mass and energy, and leave it at that. But both prehistoric and modern observers would grant you that it has probably not always been as we see it today. Time works changes in things and it is difficult to understand. In the modern world we complain that there is never enough time, and our view of it is different from that of our ancestors, as the disciples of Einstein and some of our modern philosophers explain.

But time is not exclusively in the bailiwick of physicists and philosophers: historians cope with the flow of a few centuries or millennia, geologists delight in eons of millions of years and astro-nomers are greedy for thousands of millions of years in which to create stars, spin galaxies and send light bouncing round the universe. Of these gentlemen the geologist is perhaps the most fortunate. He is not much concerned with the follies and foibles of mankind nor with the intangible immensities of the cosmos: he is interested in the earth beneath us, how it came to be as it is and how it has provided the stage for life itself. His interest, as a pioneer geologist Charles Lyell wrote in 1830, 'is that science which investigates the successive changes that have taken place in the organic and inorganic kingdoms of nature...'

To many an early Victorian scientist, and Lyell knew a large number of them, nature was a very orderly and rather precise matron.

The idea that she has the long and eventful history that geologists hinted at was novel, daring and even revolutionary. Charles Lyell was the first professor of a new subject at the University of London, and his subject, as he put it, enquired into the causes of the changes in the organic and inorganic kingdoms of nature and the influence those changes have exerted over the years. He was a man destined to change his subject and was himself to be fascinated by the idea that geology is concerned not with dead, dusty things, so much as with the vital, dynamic and unceasing changes in the face of the earth. His view of the subject is with us still, and in recent years it has been emphasised by the glimpses of the earth from within (so to speak) offered by modern geophysics, and from without provided by the Apollo space programme. Geology has come a long way since Lyell's day but it nevertheless remains a rather recently developed body of knowledge about the planet on which we live. Geologists think of themselves as followers of a distinct discipline, calling upon other disciplines to help work out the problems it sets them. In the fashion of scientists, geologists have proliferated into different branches of the geological family but they are all 'earth scientists', even if they can scarcely understand one another's jargon. The great leaps forward in geology take place when someone provides a new unifying thought or theory. Geology recently had one, and its effects are still rumbling round the laboratories and lecture rooms. The theory of continental drift and 'plate tectonics', as the two linked concepts are called, is likely to be the major contribution of the century to geology. It concerns the evolution of the surface of the earth and implies that our old planet is still full of vigour.

Geologists are usually explorers at heart, eager to know the nature of mountain tops and oceanic deeps. Theories that embrace the birth, evolution and death of the desolate places of the world will appeal to them. Streams, glaciers, and the restless sea figure in their studies, and the results of space research are eagerly scanned for data against which to test the new ideas. Each step forward, the geologist hopes, will give him a better view backwards, into the past, to reconstruct what has transpired within the solar system and to unravel what time has wrought upon this particular planet.

Although there may be other worlds with intelligent beings in the universe, there seems to be little doubt that this planet is the only one to be so inhabited within the solar system. Earth is unique in the system: it is simply the wet planet. It has an atmosphere of gases and vapours swirling about it *and* it has a hydrosphere, a film of water turbulent and fitful. The two react, circulate and attack the outer shell of the solid earth with tremendous vigour and they seem to have been doing so for three and a half thousand million years or more. Such activity accompanied the evolution of what may be another unique phenomenon in our system—the biosphere, the complex of affairs we call life.

This book takes a rapid look at how the evolution of the solid earth—the geosphere, and its attendant atmosphere and hydrosphere together with the biosphere, earth's thin film of living things—has progressed in the four and a half thousand million years or more since the earth began to cruise around the sun.

A Little Background

As our knowledge of the surface of the earth has increased so has the need for geology to enlarge its scope. Today's sophisticated, computerised and satellite-aided science is a far cry from the small compass of the early Greek natural philosophers. Nevertheless something of the same old reverence for, and attitude to, knowledge persists, and the style of thinking has not undergone as much of a change as the passage of a few centuries might suggest. Although there are earth scientists who might scarcely know which end of a geological hammer to hold, they recognise that the basis of their subject has been established by men whose only tools were as simple as a hammer and a magnetic compass.

The Greek astronomers preferred to regard the earth as round, rather than flat. Their wisdom was lost throughout the long dark ages and not until the seventeenth century was our knowledge of the solar system and the laws of nature good enough to permit even the earth's mass to be determined. Meanwhile the conquest of materials had been progressing. For industry, war, and art, metallic ores, building materials and decorative stones were sought and used. Engineers perforce had to learn something of the nature of geological formations. Coal, petroleum and salts were mined or collected in greater quantities as civilisation progressed. In the halls of learning and in the curiosity cabinets of the royal houses of Europe, collections of minerals, rocks and fossils were amassed. In time many of these objects were described in the all-embracing treatises on agriculture or metaphysics and theology that were so beautifully produced and highly valued in the seventeenth and eighteenth centuries. Many of these folios had a special chapter on those mysterious works of the devil or of the 'plastic force' in the rocks, fossils.

The Renaissance had indeed brought some progress in natural science and interest in scientific matters grew in every country in Europe. The Royal Society was founded by Charles II in London and included papers on geological topics among its first proceedings. In the nineteenth century there were many national and local societies devoted to natural history, science, arts and letters. In universities, clubs and salons talks were given to a widening range of interested people. The impact of many of the new ideas about science was steadily growing. The Geological Society of London, the first body devoted to the earth sciences, was born in 1807: a hundred and fifty years later the number of recognised geological associations ran into hundreds.

Throughout these years there were very few people who could be described as professional geologists. The situation is very different today and the second half of our century is proving to be as exciting and rewarding a time as any that science has known. More scientists, devoting their working days and nights to investigating the nature of things, are at work today than in all the previous centuries put together.

One of the consequences of such a bloom of scientific research is an apparent fragmenting into specialist disciplines. One kind of scientist may find it hard to understand what another is talking about. Geology has been fairly lucky in this respect, but the mineralogist may have some trouble understanding the jargon of the student of fossils (the palaeontologist) and so on. We have geo-chemistry, geophysics, and other realms of study all of which are aspects of earth science and cannot separately exist without one another.

Each specialist is pleased by new discoveries in his own field, but all specialists experience a rather different kind of satisfaction when new ideas or concepts help relate hitherto separated studies and researches. Geologists have always been aware of this need for unifying theories, and the 'Father of Geology' set them on the right path as far back as 1785. He was James Hutton, an Edinburgh physician who presented to the public his *Theory of the Earth* in that year.

It was quite a revolutionary thesis, not very bulky, but it sought to explain the geological features we see around us by reference to natural processes that are also observable today. Geologically, Scotland is a complicated country and it was only to be expected that the range of rocks and minerals should attract the attention of observant people. Hutton was impressed by the kind of story that he felt could be told from the rocks of eastern Scotland. It must be, he felt, a long and eventful history, but perfectly explicable in terms of what is happening around the world now. In 1795 his work appeared in book form but it never made the mark upon the reading public that one might have expected. Hutton's style made rather stiff reading. Soon upon the scene was John Playfair, who admirably championed and illustrated his friend Hutton's ideas in a most readable volume in

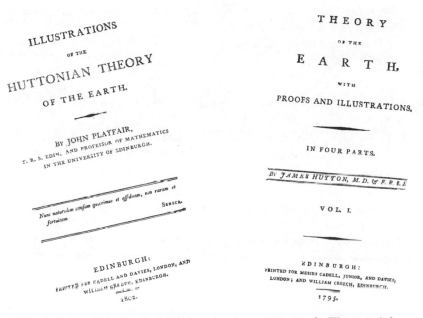

Figure 1-1 Starting posts: The title pages of Hutton's *Theory of the Earth* and Playfair's *Illustrations of the Huttonian Theory*, two texts which were to set geological thought moving in Napoleonic times and which have influenced the development of earth sciences ever since.

1802. Hutton's dictum that 'the present is the key to the past' was shown to give us a working rule of thumb which stood up to stiff criticism. At this very beginning it was made clear that geology is in essence a historical science, concerned with trains of events that require long periods of time.

Not many years passed before there was a strong feeling in some quarters that these new ideas about the origins of much of the physical world were in conflict with the biblical record and the teaching of the church. The revolutionaries in France had welcomed the new science of geology, so it must be suspect.

When Charles Darwin published his monumental survey of the living world and his ideas on evolution in his *Origin of Species* in 1859 the ecclesiastical roof fell about his ears. A confrontation between church and science had to come. Darwin's friend, Charles Lyell, had suggested on the evidence of fossils that very long periods of time had been necessary to accomplish all that could be found in the record of the rocks. Bishop Ussher's 4004 BC date for creation was not geologically supportable.

Before long there was a lot of thought given to this problem of how

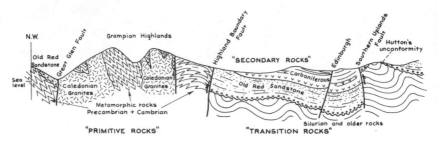

Figure 1-2 Hutton's stamping-ground was eastern and central Scotland where volcanic rocks, flat-lying and disturbed sedimentary strata, and deformed metamorphic rocks are all to be seen. It was at the unconformity at Siccar Point that Hutton realised how the geological record is locally made up of episodes of rock formation and episodes of rock destruction.

much time was implied by the geological theories. Several geologists and physicists, including one of Britain's most remarkable men of science W. T. Thompson (later Lord Kelvin), thought it might be 60 million or even as much as 400 million years. Darwin had shown that life had developed continuously since it first appeared on earth, relating to and responding to its environment and changing as the environment changed. His theory was attractive, too, in that it enabled biologists to explain the immense variety of living things and to suggest ways by which this variety had been achieved. The voyage round the world in H.M.S. *Beagle* had given him an insight into the kinds of problems we experience when trying to reconstruct the appearance of living things from the past. Darwin knew that science was far from understanding how fossils, from oldest to youngest, had come down through time to us, yet he suggested that at least 140 million years might have been needed to bring about all the changes that have taken place since the oldest and most primitive fossils that he knew had been living creatures.

Darwin's work proved to be the greatest single stimulus ever given to biology, but the succession of fossils in the rocks still offered many puzzles. The distribution of plants and animals in the geological past was clearly very different from that of today. To give one common example, fossil corals are present in rocks at latitudes far higher than those in which corals flourish today. The rocks themselves, now forming parts. of hills or mountains, were attributable to the compression and hardening of sediments deposited on the sea floor. Great changes in climate and geography seem to be indicated by the present distribution of such rocks. Upon these changes hinged the fate of the animals and plants that lived where the original sediments were laid down. As the changes took place, life evolved to take advantage of the new conditions. All this had in fact been recognised by Hutton. He saw that change in the earth itself was, and is, continuous and that it proceeds slowly by comparison with organic evolution. Science during Victoria's long reign tended to confirm these ideas.

Somehow, geology, like most other branches of science, developed in a rather piecemeal way in the nineteenth and early twentieth centuries. At first it flourished most in western Europe and on the eastern seaboard of North America, but with the expansion westwards in the United States and Canada North American geology played an increasing role in the science as a whole. It was only natural that the geological features of these regions should contribute largely to the development of geological thought. From the splendidly fossiliferous rocks in the cliffs and mountains there geologists could in relatively few years piece together the outlines of geological history. The folded nature of many of the bedded rocks and the wide variety of formations that are now crystalline, but which were once molten, were interpreted as signs of unrest and movement in the crust of the earth.

The geological story pieced together from such clues was regarded as falling into a number of recognisable chapters. Each chapter is founded upon a number of easily identified rock formations. These were produced mostly from the compression of layers of sediment deposited on the sea floor, and within the sediment lay the shells and remains of creatures or plants that lived in the sea or not far from it. As the land was worn by weathering and erosion the sediments were borne away to the sea. Layer upon layer took shape as time

progressed and each one had the mark or stamp of some characteristic of the area and time in which it was formed. Obviously the oldest
layers, the first formed, occur lowest in the succession; the youngest
are at the top.

In the course of millions of years the thickness of strata formed by
the hardening of these layers has been very great. It amounts to
thousands of metres, but because of the restless nature of our planet
the strata do not occur as a complete and unbroken column in any
one place. Periods of erosion intervene between times when sediment
is accumulating on the sea floor or elsewhere.

The strata between the breaks provided by these erosional intervals
were called systems and each was held to represent a separate period
of geological time. In the course of 180 years or so the stratigraphical
table based upon these concepts has taken shape. It is still worth
regarding each system as a document which records a chapter in
earth history and the stratigraphic column is the list or table of
contents of this history. The sub-sections and divisions of the chapters
have become more refined in recent years and the pages constantly
reveal new characters in the plot.

Studies in the Alps and other mountain chains of Europe and in the
Appalachians of North America helped mould early ideas about the
geological evolution of these continents. In time the geology of the
western cordilleras of the Americas and of other mountains fostered
new hypotheses about mountain-building and the transformation of
sedimentary rocks into hard, new crystalline materials. From the very
outset of geology it was believed that somehow there was a cycle or
rhythm in the behaviour of our planet. The erosion of the lands
produced sediment. Sediment became sedimentary rock through long
burial and slow chemical change. In time the buried rocks were
squeezed and heated by the interior forces of the earth. Many are so
affected, so metamorphosed, that to all intents and purposes they melt
and even assume some of the characteristics of the molten volcanic
material of the earth. Heaved up into mountains or continental
uplands they are once more exposed to erosion and break down into
particles that will form sedimentary deposits.

This rather simplified outline of the geological cycle has fascinated
geologists since the days of Hutton. There are short cuts across it and
it has operated rather differently from one region to another, but the
geological cycle—or rock cycle—seems indeed to have been slowly
moving round for thousands of millions of years. To keep it going
there has to be a supply of energy. It comes from two sources, from

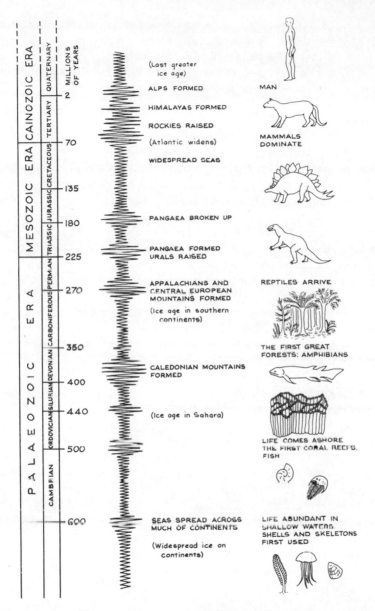

Figure 1–3   The last chapters in the evolution of the earth have occupied a mere 600 million years, but in many ways they are the most easily understood and, with their included fossils, they are literally the most vital in our story.

the sun and from the interior of the earth. The sun heats the surface and causes the atmosphere and seas to circulate, while the interior heat of the earth keeps the crust in ceaseless movement, slow but irresistible. The history of the crust of the earth is largely a story of how these forces have been spread in time and space.

Heresy!

No geologist ever seems seriously to have challenged the idea of the geological cycle. From so many parts of the world there is clear evidence that it works. And as the exploration of the continents has gone on the geography of past times has been reconstructed on wider and wider canvasses. The rock cycle brought about the changes from time to time.

Until early in this century the maps drawn to show where land and sea lay in the distant past outline very different features from those of the world today. An increasing body of evidence had suggested that in the past land links between our separate continents must have been firm and broad. At other times there may have been barriers to prevent the migration of organisms from one area to another seemingly close by. Admittedly the strange maps that geologists had to draw to explain much of the distribution of fossils and rock formations only concern a small late fraction of earth history. To many geologists they never really seemed convincingly to represent how the crust had behaved. No proof of another kind of behaviour was available, but the sceptics were restless.

Then in 1912 a young German scientist, Alfred Wegener, suggested that the continents had drifted to their present positions from the break-up of a single primaeval super-continent. No great land bridges for him: the geology of the continents was to be explained in a different way. Although he was not the first to make such a suggestion, Wegener drew evidence from so wide a field that his thesis made an immediate and sharp impact upon geologists.

The fragmentation of a single vast continent, *Pangaea* (from the Greek for *all* and *earth*), came according to Wegener at the end of the Mesozoic era. Wegener's notion implied greater changes to the face of the earth than had been contemplated for periods earlier than the Cretaceous. It was not very readily accepted by the scientific world. While Wegener was being particularly impressed with the similarities of the opposing sides of the south Atlantic, the contemporary Austrian geologist Eduard Suess felt compelled to suggest that Africa, South America, India and Australia were once part of a gigantic

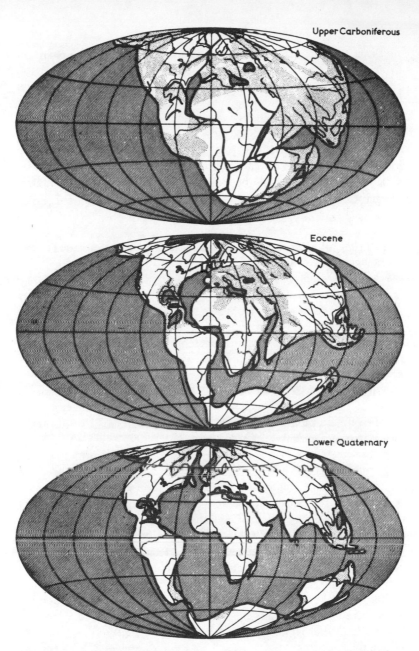

Upper Carboniferous

Eocene

Lower Quaternary

Figure 1–4   Wegener's maps of the progressive break-up of *Pangaea* have for geologists become something of a classic work. Scorned for several decades, their basic truth now seems to be demonstrated.

(*From* The Origin of Continents and Oceans, *by courtesy of Methuen & Co.*)

southern continent. He called it *Gondwanaland* after Gondwana in
Central India.

Although Wegener had an able champion in the South African
geologist Alex du Toit in the 1920s and 1930s, his theory of
continental drift was shunned by many geologists and most geophysicists. They could not see how continents could be transported
sideways, and they found objections to almost every aspect of the
idea. The controversy spread over some forty years: the idea had its
attractions but to prove the substance of it was another matter
altogether.

Revolution

The theory of continental drift gave geologists a completely new
framework upon which to build their reconstructions of the earth in
the not-too-distant past. Despite the continuing objections of some
geophysicists and even of some geologists, it has become widely
accepted in recent years. Ironically, it has been largely because of
the spectacular results of work on the physics of the earth that
continental drift has acquired respectability. In the geologist's imagination there is now an entirely new model of how the earth is
constructed and how it 'works'.

While it has long been accepted that big vertical movements in
the crust of the earth can take place, the image of continents carried to
and fro, for example, on currents within plastic layers beneath them
has been rather hard to accept.

Within the space of a very few years the whole situation has been
turned about. Today not only can a case for continental drift be made
but a mechanism, seemingly in accord with the geological facts, can
be proposed. It means that the history of the earth—or at least of its
continents and the oceans between—can be reviewed in a very
different light from that of a decade ago. The movement of continents
may have climatic and geographical effects which in turn may
clearly relate to the evolution and fortunes of life itself on this planet.

Three major advances took place in geology in the 1950s and
1960s, to make this rewriting of much of earth history possible.
They were techniques that enable us (i) to explore the hidden ocean
floors with a thoroughness never attained before, (ii) to measure with
reasonable accuracy the ages of rocks and minerals, and (iii) to
measure the magnetic properties that were imparted to rocks when
they were first formed. To these a fourth should perhaps be
added—the means of analysing the global quivering that takes place

when major earthquakes occur. Together these new developments led to some serious questioning of the nature of the ocean floors between the continents and the answers that were made led to a new and exciting hypothesis, known conventionally as 'plate tectonics'. This revolutionary idea suggests that the crust of the earth consists of a series of slowly moving slabs or plates. The plates originate from the upwelling of volcanic rocks, basalts, along the rifts discovered in the great ridges beneath the oceans of the world; they disappear at other places, ocean trenches for example, where they are drawn down into the deeper layers. The continents sit upon some of the larger plates and the growth, movement and destruction of these plates may account for continental drift and for the great episodes of earth movements, mountain-building and vulcanicity that have marked geological time from era to era.

The search for evidence of these 'plate' movements goes on and

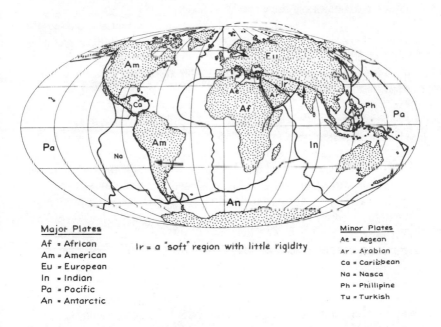

Major Plates		Minor Plates
Af = African	Ir = a "soft" region with little rigidity	Ae = Aegean
Am = American		Ar = Arabian
Eu = European		Ca = Caribbean
In = Indian		Na = Nasca
Pa = Pacific		Ph = Phillipine
An = Antarctic		Tu = Turkish

Figure 1-5 The cracked shell of the earth is made up of six major 'plates' and six or seven minor 'plates'. Movement which can now be detected between adjacent plates of crust seems to be a major force in controlling the shape and positioning of the continents and oceans.

many geologists are actively trying to relate 'drift' and other earth-movements to the structure of the ocean floors. Ocean-going research ships from many nations are taking part in these adventures, but as a contribution to an International Decade of Ocean Exploration rather than specially to investigate aspects of the local ocean floor geology. Even more promising in terms of results for the geologist and geophysicist is the Joint Oceanographic Institutions Deep Earth Sampling programme—JOIDES for short. The aim of this project is to drill a series of boreholes in the ocean floor. To do this special ships have been designed, capable of maintaining a fixed position over the ocean floor and carrying out the kind of operation normally used in the search for oil. If the ocean floor is growing from the ocean ridges, older sediments and floor should be found at the edges of the oceans than near the ridges. The JOIDES samples provide the test of this and so far the *Glomar Challenger,* the ship commissioned and fitted out to do the major work, has completed 24 special 'legs' or drilling cruises sampling the floors of the world's oceans.

From this and the work of other teams of scientists around the world there is an ever-increasing flow of data, and a growing understanding of how the planet works. Knowing better how the machine ticks at present, we may be able to deduce more accurately what its performance has been in the past.

Earth Matter

The three states of matter, solid, liquid and gas, have a complex distribution in our planet. Customarily we think of the earth beneath us as solid, while round about flows, trickles or laps the hydrosphere, fanned or stirred by the blowing movement of the atmosphere. The earth is visibly composed of rock, which at the surface may decompose to produce soil, sand and mud. Many of these superficial deposits are complicated mixtures of different substances derived from the reaction of the rock materials with air and water. A knowledge of the nature and origin of rocks is fundamental to understanding the earth, and since rocks are composed of minerals we need to know something of them too. Recently developed rapid and accurate methods of geochemical analysis have accelerated our study not only of the materials themselves but also of the processes in which they are involved.

The hydrosphere includes all the water in the planet and provides an almost universal solvent and a powerful medium for moving other materials around. Without water many of the chemical reactions that are an essential part of geological processes could not take place, transport of weathered rock material or sediment would be infinitely less effective and life, as it has developed on this planet, could not exist.

The atmosphere is another part of what we might call 'the geological cycle'. It contains two geologically and biologically important gases, oxygen and carbon dioxide, as well as a lot of nitrogen and traces of several relatively inactive gases. One of the conspicuous functions of the atmospheric envelope around the earth is to spread the effects of heating and cooling between the tropics and the poles. The result is a never-ceasing process of change which we call weather, or, more broadly, climate, and the effects of this are obviously all-pervading. In another way too the atmosphere is important; it helps to shield life on the surface of the planet from the bombardment of cosmic particles from outer space, and to filter out the biologically harmful ultra-violet radiation of the sun.

All in all, however, we can say little about the composition of the earth beneath us that is based on direct observation. The atmosphere and hydrosphere can be sampled relatively easily but only the topmost eight or nine kilometres of the solid earth have been penetrated by boreholes or mines. Geochemical studies of certain ancient and volcanic rocks may provide information about the deeper parts of the crust and of the immediately underlying zone, but it is relatively sparse. Nevertheless, geophysical investigations can suggest the composition of the regions within, but we may leave these till a later chapter.

At first sight the rocks exposed as the surface of the continents seem to be so varied that they may be unreliable sources of information about the overall composition of the crust. However, even allowing for the formation of new crust by volcanic activity, it may be not unreasonable to assume that most if not all the rock formations have been derived in the long run from the first formed rocks. The original crust we believe to have been made of *igneous* rocks: that is rocks solidified from a hot molten state. So if the mean composition of the igneous rocks known at the surface can be calculated, an idea of the average composition of the entire crust may be achieved.

In terms of common rock compositions it may be said that the continental crust is intermediate between that of the two commonest igneous rocks, granite and basalt. The most abundant element present is oxygen, occupying some 46·5 per cent of the earth's crust: silicon is second. On a volume basis oxygen may be as much as 94 per cent of the crust since the oxygen atom is so much larger than the atoms of silicon, aluminium, magnesium, or the other most common elements.

Our knowledge of the composition of the solid earth comes from the study of the constituents of the rocks themselves, the minerals. Mineralogy is concerned with the naturally occurring, mainly crystalline, chemical compounds of fixed or only slightly varying composition. Aggregates of minerals produce rocks—*igneous* from the molten state, *sedimentary* from the weathered remains of previous rocks, and *metamorphic* formed by the action of heat and pressure on existing rocks. Although the study of minerals as objects of interest in themselves dates back to classical or even prehistoric times, the study of the nature and origin of rocks—as distinct from their distribution—is a comparatively recent development. In the mid-nineteenth century H. C. Sorby ground the first slices of rock thin enough to be examined by transmitted light under the microscope, and he thus opened up an immensely productive field of research. The microscopic characteristics of many rocks were revealed for the first time and something of the full range of rock composition and texture was apparent. The technique has not been changed very much in the years since then. One important outcome was the realisation that most rocks are composed of a comparatively small number of minerals, the majority of which are complex silicates, the so-called 'rock-forming minerals'. More recently has been developed the application of analysis using beams of electrons in place of light and the experimental investigation of mineral and rock compositions at high temperatures and pressures. X-ray techniques have been devised to reveal the internal structures of crystals. Chemical composition largely controls the type of structure present, so by using X-rays we have an aid to identifying mineral composition. The study of minerals has contributed in no small way to an understanding of the (atomic) structure and behaviour of crystals and, indeed, of matter itself.

The silicates stand out as the most important group of rock-forming minerals. They all incorporate the key atomic grouping of silicon and oxygen, the SiO_4 unit, sometimes called the silicon tetrahedron. It is a strongly electromagnetically bonded quartet of oxygen atoms around a much smaller silicon atom. In the silicate minerals these units may be scattered independently or may be linked in various ways with the metallic ions. Chains and sheets of tetrahedra interspersed with arrangements of ions of other elements are basic to the structure of most rock-forming minerals.

Many of the rocks encountered by the eighteenth-century European scientists are visibly composed of silicate crystals, together with minor quantities of other minerals. To account for these rocks the hypothesis was put forward that they were the chemical precipitates of a universal primeval ocean. It was not long, however, before criticism of this idea produced quite heated argument in the lecture rooms, salons and schools. This 'neptunist' concept of oceanic precipitation eventually gave way to the views of the 'plutonists' who could show that many igneous rocks were not aquatic precipitates but had solidified from hot mobile fluids (*magma*), originating deep within the crust of the earth. Volcanic activity provides a view of such fluids reaching the surface and the 'neptunist' scheme was conclusively demolished by James Hutton, whom we have already mentioned. Hutton recognised ancient volcanic rocks when he saw them, and his ideas were well in accord with observable fact.

During the century that followed the publication of Hutton's *Theory of the Earth* geologists described a wide range of igneous rocks from many parts of the world. They recognised that the majority of intruded igneous rocks (not breaking through the crust) are like granite in composition, rich in quartz and the aluminium and alkali silicates known as feldspars. Most extruded rocks—the subaerial or submarine products of volcanoes—have a basaltic composition with abundant iron-magnesium (ferro-magnesian silicate) minerals. Between the quartz-rich light-coloured granites and the quartz-poor dark basalts exists a wide range of rocks of intermediate chemical composition. Moreover there are influences which may determine the kinds of minerals produced when the liquid crystallises. In a subterranean pool of molten rock (*magma*) early-formed dense crystals may sink, leaving the upper reaches of the body different in composition from the lower part where the crystals settle. Experiments by N. L. Bowen in the Geophysical Laboratory in Washington in 1915 showed that this process can operate in nature. Since then geologists have described from Greenland, North America and elsewhere, igneous masses with a layering of the densest crystals at the bottom.

Ever since the first igneous rocks solidified at the surface of the earth there have been processes of destruction at work on them. Changes of temperature impart fractures and initiate mechanical weathering. In reaction with water and the atmosphere, chemical weathering of the rocks may produce new minerals from old. Most silicates may be affected in this way, particularly if the water is charged with carbon dioxide, which is common enough. Virtually all the rock-forming minerals, except quartz which is simply silicon dioxide, break down comparatively readily under chemical weathering. Clay minerals and soluble carbonates are born in this transformation, and together with quartz, they form the bulk of the sediments that accumulate on land and in the sea to produce the layered sedimentary rocks. Most sedimentary rocks consist of such particles of pre-existing minerals and rocks, but chemical precipitation figures conspicuously in the origin of some of them. Organic activity, too, is instrumental in producing enormous volumes of carbonate rock in tropical regions. Carbonaceous deposits such as coal also are the direct products of the activity of life. Henry Sorby turned his attention also to the study of sedimentary rocks and devised experiments to show how many of them may be produced in nature. He attempted to reproduce some of the processes through which sediment would pass on its way to becoming a sedimentary rock. Although they may bring about marked changes in the sedimentary material, these processes take place at the relatively low temperatures and pressures which prevail at or near the surface of the earth.

At greater depths, however, or in the proximity of magmatic intrusions or a volcano, the changes brought about by heat and pressure may continue to such an extent that a complete reorganisation of the mineralogical composition of the rock takes place. In a continuous process of change or metamorphism the whole character of the rock is altered. A suggestion popular a few years ago was that the ultimate product of metamorphism on the regional scale was granite: in some cases the process was assisted by the migration of hot fluids and gases into the metamorphosing terrain from below rather than by simple pressure and melting. This 'granitisation theory' had many attractive applications to individual granite and metamorphic rock complexes. Today, however, the simpler theory of differential melting under high pressure seems adequately to explain the complex relationships of many granite bodies and the somewhat less 'granitic' rocks known as *migmatites*. Although the mechanism of transformation may not be quite what the 'granitisers' had in mind, their dictum '*Per migma ad magma*' seems to hold true.

Geological Time
Hutton's 'Law of Uniformitarianism' presupposed unheard of periods of time to accomplish the changes and events indicated in the geological column. The biblical deluge was discredited but the nature

Figure 1-6 Humorously shown in Dr Gilbert Wilson's cartoon is the rock cycle – the unending progress of material in the crust from its volcanic source, through weathering and deposition to sediments and sedimentary rock, ultimately becoming metamorphosed and remelted deep in the earth.

(Reproduced by courtesy of the Council of the Geologists' Association.)

of what should take its place varied from one authority to another. William 'Strata' Smith and other practical geologists demonstrated in the early nineteenth century that the correlation of strata could be made by recognising the fossils within them. As the sequences in which assemblages of fossils occur were slowly pieced together one above another, a relative time scale of their ages was built up. Thus in Britain 'Strata' Smith produced a table of rock formations showing the order in which they occurred in the ground, the oldest at the bottom, the youngest at the top. This stratigraphic column represents all the deposits formed during the last few eras of geological time, and the separate identities of recognisable periods of time based upon the rocks listed in the column were eventually established. The names have changed relatively little over the years and it says much for geologists' willingness to co-operate with one another that most of these stratigraphic names are used internationally. Discussion of what the fossiliferous rock succession means in terms of years was throughout the last century a heated and sometimes acrimonious business. Even the most outrageous of these estimates of the span of time between the earliest Cambrian and the present, however, would seem inadequate by modern calculations. Various methods were used to determine this span, to quantify the time scale.

Attempts to use the rate of increase in the saltiness of the sea and the rates at which sediments accumulate ran into great difficulties, and estimates of geological time ranged between 20 and 100 million years. Charles Lyell, ever alive to the importance of the issue and to the possible use of the newly formulated ideas of organic evolution, hazarded an estimate based upon assumed rates of organic change. He came surprisingly near the mark. Obviously only processes which continue in a single recognisable way and which have left tangible testimony in the rocks can be used, and organic evolution is one of them. Even in Lyell's day, however, it was granted that the fossiliferous geological column represents only a fraction of the total amount of time that has passed since the geological record began.

Sooner or later physicists were bound to take an interest in the problem of geological time. It has an important bearing on their concepts of the origin and evolution of matter. In 1897 Lord Kelvin calculated the age of the earth on the assumption that it had cooled from an initial molten state. His several ventures into this field also considered the origin of the sun's heat, and indicated an age of between 200 and 400 million years for the earth. While in some ways this was an advance, Kelvin's authority and sophisticated mathematical approach unfortunately persuaded some geologists to condense earth history into this rather short period of time. They preferred his theory to their own reasoning, and it was not the only occasion on which geologists were to defer to the views of the pundits of physics.

Very shortly after Lord Kelvin's work was published other developments in physics were to be felt in geology. The discovery of radioactivity showed that some elements such as uranium were unstable, decaying to produce other varieties of the elements known as radio isotopes. The rate of decay by radioactivity of some elements can be measured and the measurement of the ratios of the 'parent' and 'daughter' elements in rocks and minerals can reveal how long they have been present in the rock. Where these elements occur in igneous rocks the radioactive products date from the time the minerals were first crystallised or the rock solidified. The realisation that this process could be used as a form of geological clock came to Lord Rutherford, who succeeded Lord Kelvin in the Cavendish Laboratory at Cambridge. He estimated 500 million years for the age of a uranium-containing mineral. The American chemist, Bertram Boltwood, at Yale University, suggested the ages of several geological periods on the basis of radiogenic lead/uranium ratios in minerals. He even suggested that minerals from Precambrian formations might be as much as 1,640 million years old. There was an immediate outcry. The idea of such large periods of time was highly suspect in the eyes of many geologists, who believed that there must be analytical errors in the calculations. However Boltwood, and the much respected Arthur Holmes in Britain, persisted in the belief that geochemical work could provide geologists with a reliable means of measuring time. The rather crude analytical techniques of the early decades of this century have been replaced by the use of a sensitive analytical instrument, the mass spectrometer, invented in the 1950s. Holmes' enthusiastic teaching and vision bore fruit in the results obtained from using these new techniques as he had advocated.

Development of the methods of radiometric age determination has continued and certainly there will be further refinements of the time scales produced. Many thousands of these determinations have been made and no doubt seems to exist that the history of this planet extends back over several thousand million years. A figure of 4,500±50 million years is now supported by estimates based on analyses not only of materials from the oldest available rocks on earth but also of meteorites and rocks brought back from the moon.

The range of radiometric methods available to the geochemist has grown and dates can now be obtained from ancient lava flows and volcanic beds, from igneous intrusions and from some sedimentary rocks. Certain restrictions have become apparent. Metamorphism may lead to isotope loss and so an earlier age is indicated: and metamorphic rocks can only be relied upon to give worthwhile dates when the relative age of the metamorphism is known.

In the early 1960s radiometric analyses from all parts of the world were available. To several geologists they suggested that, despite the rather large range of ages possible for any rock sample, there was a tendency for the determinations to group at certain times in the geological calendar. Naturally, there is less certainty about the reliability of the higher figures, but there seems to be no denying that, with few exceptions, the oldest rocks available are about 3,600±200 million years old and that 'clusters' of radiometric dates do occur. John Sutton of Imperial College, London, has suggested that major phases of deep crustal activity with the production of large volumes of

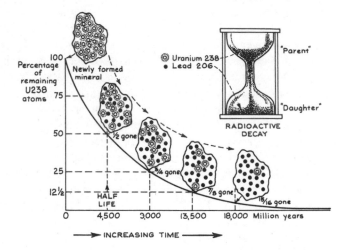

Figure 1-7   The radioactive clock is illustrated by the decay of uranium 238 to lead 206. By radioactive decay the parent material loses half of its atoms over a period of 4,500 million years. This figure is known as the half-life and the long half-life of uranium 238 makes it useful to geologists as a form of clock recording the lapse of time since the parent mineral was first crystallised. (After Fagan.)

deep-seated, or plutonic, rocks may have occurred during these periods. We shall return to this idea in Chapter 5.

This may be helpful to the geologist concerned with the most ancient rocks, but to the palaeontologist a more reliable dating of events in the history of life would be more welcome. Unfortunately there are few radiometric dates that can be correlated to particular fossiliferous beds and hence to events in organic evolution.

Nevertheless, radiometric dating has been of great service to investigations into the time and duration of events within the evolution of a mountain belt and into the evolution of the continents, and in testing the theory of continental drift, to name but the most general.

It confirms that only a small part of the history of the planet is recorded by the fossiliferous rocks—or at least the rocks younger than the Precambrian systems. This span of time, about 600 million years, is known as the Phanerozoic eon; the earlier span, the Cryptozoic eon, is seven or eight times as long.

Earth Evolution

Evolution is the process of change, the activity of becoming something new and different from what existed before. Most of us associate the word with the history of life and the changes that have taken place in the biological world throughout time. Organic evolution, despite its many puzzling features, is regarded as a proven fact, and it has been the dominating concern of biological research in the first half of the twentieth century.

The background to evolutionary biology, however, has been the evolution of the earth itself. While evolution has occupied the minds of biologists, the majority of geologists have been compiling details about the regional variations in the surface rocks of the earth. From these labours, which are still far from complete, patterns of earth movements in both time and space have become apparent in the geology of the continents. These patterns are locally reflected in the kinds of sedimentation that took place—thin uniform accumulations with many limestones and precipitates on the flooded margins of the continents; coarse gravels and sandy deposits within the upland regions, and thick piles of predominantly silty or muddy sediments in the deeper water areas along the continental edges. A burgeoning interest in sedimentology in the second quarter of this century has enormously increased our understanding of how sedimentary rocks originate and (hence) our ability to interpret the ancient environments they represent. Here is a branch of geology in which the present seems to be a very effective key to the past. But it has also become apparent that movements taking place within the crust have been very important in bringing about the major features of the earth's surface upon which these ancient environments were founded. Structural geology has in no way lagged behind other branches of earth science in the last few decades. In the structures of the Alps, the Western Cordillera and other mountain ranges today we can see the effects of relatively recent, if not actually continuing, earth-movements. The ancient terrains of the continental interiors, the shields as they are sometimes called, reveal structures that are akin to those of modern mountains but many times more antique and deeply eroded.

All in all, Hutton's uniformitarianism seems to work well enough in unravelling much of the history of the Phanerozoic eon. Perhaps it is far from being as universally applicable as was once thought, but it does not seem to involve us in severe dilemmas. More than one shrewd geologist has remarked that in its way the past is the key to the present, a perspective against which to measure today's natural world.

With the seven-eighths of geological time that constitute the Cryptozoic, however, we cannot be so sure that a Huttonian view of things will serve as well. The oldest rocks yet known, $3,980\pm170$ million years old, occur in west Greenland and they reveal features that are not seen in any younger formations. They seem to have experienced conditions which have not commonly occurred since. There are many other features in other ancient rocks also that indicate conditions far different from those either prevailing today or suggested in the rocks of the Phanerozoic eon. Our means of interpreting them however are decidedly limited.

Clearly earth has evolved. If it is as much as 4,600 million years old, as seems to be probable from the data on radiogenic elements, of the first 500 or more million years the rocks tell us nothing. From only a very few rock formations can we glean anything about the next 500 million years or so, and much of what we suggest is tentative or mere speculation. The next two thousand million years have left a somewhat less tattered record, and for the remainder the evidence gets progressively a little better. It is the geologist's profound misfortune that as he traces the record of earth evolution further back in time he finds conditions less like those he understands and the record more scrappy and blurred.

To picture the birth of our planet we have to look at the stars and the evolution of solar systems in progress elsewhere in the universe. To reconstruct its early evolution we have little astronomy to help us and even less geology. For almost a thousand million years or more earth's voyage through time left few decipherable features that remain to us today. As it has spun and swung round the sun incessantly since its birth, however, it has been changing, expending energy drawn from the parent sun and from its own internal heat. The changes have involved all parts of the planet; its outer gaseous and watery envelope and its fragile crust as well as its interior.

Just as the wrinkles on cooling jam or in the skin of an apple mark changes taking place well below as well as at the surface, so the events at the surface of the earth may reflect, at least in part, changes deep below. The most important changes may have involved a thickening crust, with less and less of the great internal energy escaping outwards as time has gone on. How or when continents first evolved is not known, nor when they began to drift across the global surface on plates of crust that were born, moved and devoured in the ocean basins. Continents have passed to and fro at least since Palaeozoic times and seem to have had a history of drift back into the Cryptozoic. They could not drift until there was a means of moving them such as we see today and there is no guarantee or even likelihood that this mechanism goes very far back into early earth history.

Nor is there any warrant to assume that earth's behaviour will continue in the same manner indefinitely. Her days are numbered and they will see perhaps a slowing down in many of the processes generated within the planet. The days will become dim as the sun dims and cold as the sun cools, but by then we shall have given up interest in such matters.

The Evolution of Life

Although we have touched upon the evolution of life before, it merits another line or two here. There is no doubt about the existence of evolution as a process, nor much doubt about the various lines of descent that lie behind modern plants and animals. To explain many of the pathways taken by organic evolution, however, the geologist has to provide an account of the geographical conditions of the past. Since organic evolution is a response to changing environment, our understanding of evolution is improved if we understand the environmental changes. Much of the fascination of earth history lies in the pictures we conjure up of strange landscapes, weird floras and bizarre animals, and without the contribution of the geologist, the student of ancient life, the palaeontologist, finds his principal actors without a stage upon which to appear.

Against its backcloth of sea, shore and upland, this stage has areas for crowds and areas for soloists. The manner in which it has been occupied by the performers is revealed by that joint branch of the earth and biological sciences known as palaeoecology. While ecology attempts to relate modern plants and animals to their surroundings, living and non-living, palaeoecology seeks to do the same for the creatures and plants of the past. Its working materials are rare and flimsy, despite their stony compositions, and the more ancient the stage, the fewer the actors, the simpler their parts and the less substantial are their remains. But at least we feel safe in assuming that they all had the same necessary driving force that organisms have today. They competed as individuals and as species for space and food.

The science of palaeoecology has followed on the heels of ecology but because its materials are more limited, its conclusions are more tentative. Statistical investigation of ancient animal communities, ranging from fossil shellfish to land-roving mammals, has yielded results which are worth considering even if they are experimentally unprovable. The composition of animal or plant communities and their relationships to assessable factors such as temperature or salinity have been taken up by specialists in the palaeontology of several different ecological systems. From these labours a great deal has been learned about the evolution of most of the major groups of animals and of some plants. In one field, however, we have not made so much progress. Explaining the major extinctions in the history of life remains almost as difficult as ever. Reasons for them are sought in ecological changes and in extra-terrestrial influences and events. A promising new line of study seems to be that of relating periodic extinctions of animals having a high rate of (oxygen) energy utilisation to variations in the concentration of oxygen in the air. Atmospheric oxygen fluctuations may have been a primary cause of animal extinctions. However, it is still questionable that all the crises in the evolution of life can be directly related to atmospheric or climatic events. The possibility is intriguing and receives further attention in later chapters.

Despite phases of widespread extinctions, life has undoubtedly become more diverse and abundant in all its forms since before the Phanerozoic eon began. We have a greater knowledge of marine species of the past than of other fossil groups, and the diversity of present-day living marine animals appears to be more than ten times greater than it was during the Palaeozoic. The American palaeontologist, J. W. Valentine, sees this increase in the diversity of species as the response to continental drift and steeper temperature gradients from the equator to the poles. As drift has proceeded marine life has adapted to an increasing number of new and locally or temporarily isolated environments. References to this idea figure in the later chapters of this book when the differences between the living world of

the Palaeozoic and that of the Mesozoic and between the Mesozoic
and the Cainozoic floras and faunas are discussed. It may explain
with some credibility much of the distribution of fossils, but it is
scarcely a provable theory in the experimental sense.

Warning

Despite the conscientious efforts of all its most ardent disciples for
many decades, in its development so far geology has tended to be an
inexact science. The individual observer may have little chance to
measure geological processes in action, since most of them are slow
by human standards. Nevertheless with time and technology the
means of rectifying this are growing. The continents may be changing
their relative positions by several centimetres per year—a difficult but
no longer impossible interval to measure across oceanic distances.
Many geological processes are slow, but geological time is of great
length. By human standards, both the earth and the time it has existed
are immensely large; only a fraction of the planet can ever be
examined in detail. The vast, hot interior with its high pressures
remains unknown, and only by indirect means can an estimate of its
nature ever be obtained.

The surface of the earth shows highly irregular but perhaps not
random distributions of rocks, oceans and continents. The properties
and characters of the rocks and minerals are variable and the
scattering of life in time and space seems always to have been
anything but uniform. Being a legacy from the last century, the
stratigraphic column is subdivided into arbitrarily designated systems
representing oddly unequal periods of geological time. The Cambrian
period may be 100 million years in length, but the Cretaceous is less
than a third of that. The record itself is continually being eroded
and destroyed; gaps and ambiguities exist in every direction. To
many investigators these are, indeed, some of the attractions of the
subject.

The historical geologist must thus be something of an optimist and
always seeking mechanisms by which changes to the surface of the
earth have been brought about. The discovery of 'plate tectonics'
affords him much that is useful, but there are still problematic areas.
A little fanciful fiction about the history of the planet does no harm,
provided that it is not passed off as proven fact. A dictum to be
observed by all earth historians is that they should be careful about
their facts and bold in the way they use them to suggest models or
images of the past. There is always more than one possible explana-

tion for a fact and there are several possible models of an evolving earth to be drawn up from the information we now have.

At the moment interpretations of earth history are being offered and scrapped with a speed never seen before. They all now have a dynamic aspect, an inbuilt recognition of the vigour with which the face of the earth has been wrinkling and changing as it has aged. The account on these pages thus may seem not to pay enough heed to the alternative interpretations of the evidence and to tell only one side of the story. Between the writing of these lines and the reading of them the ideas they report may no longer be tenable. What seems a certainty today may appear as myth and fallacy tomorrow. We can but tell to the best of our ability a story that is logical, possible, and in accord with the proven facts.

The kind of warning that had favour among Victorian authors might well be included here. Because the evidence for much of our history of the earth is hard come by we tend perhaps to regard it too highly. Accordingly, dear reader, in what follows for 'certainly' read 'probably', for 'probably' read 'possibly', for 'possibly' read 'conceivably', and bear in mind that in its uncertainties and paradoxes lies much of the fascination of the history of the earth.

CHAPTER II

The Face of the Earth

WE NO longer doubt that, by human standards, earth is immensely old, or that it is a complex, dynamic body travelling on a fairly regular path in space. To understand how it manages to hold its course is rather beyond the scope of this book but we are concerned to know something of the events that have taken place while it has been making its long journey. Most of the clues to these events are to be found on or just below the surface of the planet and we should start by taking a look at the complex physiognomy of earth. The wrinkles and scars it bears all have a significance. To read a man's character from his face is notoriously difficult, and to read earth's history from its thin outer crust alone is at least as tricky.

On 20 September 1519, Ferdinand Magellan set out from Seville in southern Spain on a voyage that was for its time as remarkable as that of the first cosmonaut in orbit. He, or rather his expedition, for Magellan died on the way, sailed round the globe. None of his crew fell off the edge of the world, no one was sucked bodily down into the realms of fire and demons below. They circumnavigated a globe and they brought home the most remarkable tales, some of which were frankly not credited by the man in the Seville street when Magellan's enterprise returned there in 1522. From then on the pace of discovery quickened; the Age of Exploration began and in the course of the next four centuries explorers sailed, sledged, flew or simply walked into almost every part of the world. Geographical exploration prompted questions about the events that have produced such a range of topographies, floras, faunas and peoples.

In Europe and North America the agricultural and industrial revolutions brought a flowering of scientific enquiry and commercial interest in the wider world. When Charles Darwin sailed round the world in H.M.S. *Beagle* science was bringing a degree of order to our knowledge of the globe. Nevertheless geology was in its infancy; only a small part of the northern hemisphere had been geologically mapped and ideas of earth evolution were rather primitive. The principal task was to survey the surface of our planet, to chart its oceans, measure its mountains and observe the changes, catastrophes and regimes in progress. Darwin was lucky: he saw volcanoes and experienced earthquakes, he was able to examine all kinds of rocks and to collect many kinds of fossils. He noted all meticulously and was proud to add to the fund of geological knowledge of the day. He saw that the Huttonian theory had to be substantiated by information about geological processes in action. Until this was done no view of the earth was comprehensive and no hypothesis of its history was more than fanciful.

It would not become us to assume that we now know exactly what the surface of the earth is like and how it reflects the structure within, but geologists are, so to speak, beginning to see the wood despite the trees. Our present theories about the origin and history of our member of the solar system may be no nearer the truth than those of a hundred years ago, but we hope that they explain more and are beginning to acquire some precision. Before we tell our version of the history of the planet, we need to survey something of what has to be explained. There is no doubt that, knowing more about the outside than the inside, we ought to start there, taking a look at earth, air and water.

As it spins on its axis and revolves about the sun, earth is merely a small member of the family of the solar system. Mercury, Mars, Venus and earth form the innermost quartet within the family and are known as the terrestrial planets. Apart from size and, perhaps, chemical composition, their similarities are not very obvious. Only recently have reliable pictures of the surfaces of the nearer neighbours Mars and Venus become possible. Mercury remains virtually incognito. Martians and Venusians may figure in science fiction but there is little evidence to show that such beings really exist, although some forms of life may possibly occur on those planets. Earth seems to be unique in having not only the right kind of chemical composition and temperatures for highly developed life to exist but also a very complex and largely wet surface on which life has evolved. The varied and changeable nature of earth's surface throughout at least part of the planet's existence has offered all kinds of environments in which life could acquire its multiplicity of forms. The evolution of the crust of the earth has set the stage for the evolution of life. It is hard to see how, had there never been any land, mammals and man could have arisen. Had there never been an oxygen-rich atmosphere there might indeed never have been any higher or vertebrate animals at all. One might say that we owe the pattern of our evolution and existence to the changing expressions on the face of the earth.

The other terrestrial planets Mars and Venus possess atmospheres, but poor Mercury is so near the sun that any gas at its surface is blasted away into space. Venus, on the other hand, is largely hidden from direct view by dense clouds of carbon dioxide, nitrogen and water vapour and ice crystals. Earth goes one better still and has not only atmosphere but also a hydrosphere, water in abundance in its various forms, vapour, liquid and ice.

Because of its size and density (5·52 grams per cubic centimetre) the earth has a force of gravity strong enough to retain the waters of the oceans and the atmosphere. The moon, our ancient satellite, has a mean density of only 3·34 grams per cubic centimetre and has in consequence a low force of gravity; it has lost any gases or liquid water that it once possessed. Lunar explorers have principally to face problems of 'weightlessness' and lack of atmosphere, and although they can take their own air and water with them they cannot immediately do anything about the low pull of the moon's gravity. Undoubtedly the differences between the physical and chemical changes that occur on the moon's surface and those that are at work on the earth are largely due to this lack of atmosphere. That is one, but not the only, reason for the lunar landscape being so different from our own and so monotonous.

With chemical and physical activity so abundantly in progress on the surface of the earth, it must be one of the most intensely active regions in our solar system. It is unique, wet, and fermenting. Yet all this activity is confined to a very thin zone or skin and the way in which it is distributed there is very uneven. Beneath the turbulent atmospheric cover rock and water are scattered in a complex pattern, one which has been constantly and visibly changing. Much of the energy that keeps the changes in motion comes of course as solar radiation from the sun; some is provided by the internal heat of the earth itself, fired in part by radioactivity; and some is provided by the omnipresent force of gravity pulling all things towards the centre of the earth, hence ensuring that the planet is almost spherical in shape. Gravity also provides a simple but most effective control of the behaviour of water; water always flows downhill, sometimes with tremendous force.

This flowing downhill of water has in fact divided the surface of the planet into two unequal parts—land and sea. A glance at the terrestrial globe cannot fail to impress one that the seas and oceans occupy more than two-thirds of the surface. Were the seas to rise, as they might with the melting of the present glaciers and icecaps, most of the coastal regions of the world and all the major ports would be flooded. The effect on human life would be catastrophic, but for other organisms it might have beneficial results.

Obvious as this division into land and sea is, it does in fact literally overlie another fundamental feature, one which has only been appreciated fully in the last few decades: the continents and ocean basins are geologically as well as topographically distinct entities. Beneath the

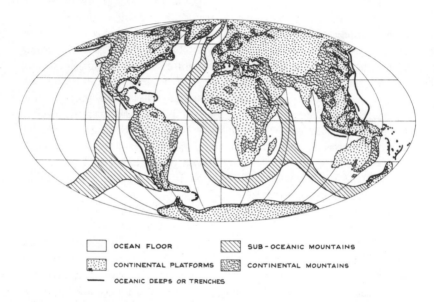

OCEAN FLOOR   SUB - OCEANIC MOUNTAINS

CONTINENTAL PLATFORMS   CONTINENTAL MOUNTAINS

—— OCEANIC DEEPS OR TRENCHES

Figure 2–1   The major surface features of the earth may be grouped under the five headings shown on the map. Each of them differs from the others in form and origin, and together they reflect the ceaseless activity of the crust.

shallow waters around the coasts the sea floor may extend some way beyond the shore before dropping rather abruptly to depths of many hundreds of metres. These 'continental shelves', covered by no more than 200 metres of water, are essentially parts of continents: their natural resources are akin to those of land areas nearby. The deep ocean floors are geologically very different, and we should look more closely at this difference before going on to examine what happens in and on continents and in and under ocean basins.

Geologists have been chipping away at the continents for many years in their attempts to find the composition of the rocks. All their evidence confirms that the continents are composed of rocks of a relatively low density, averaging about 2·8 grams per cubic centimetre. The overall composition, determined from the mean of many thousands of mineralogical and chemical analyses, is similar to that of the common rocks, granite and granodiorite. It is notable that rocks of this density or composition are exceedingly rare beneath the oceans.

Granites and granodiorites are essentially composed of the minerals quartz, feldspars and mica, with small quantities of a few other dark minerals. They are by weight about 65 to 70 per cent of the oxide of silicon, known as silica, with the oxide of aluminium, known as alumina, being most of the remainder. To denote these continental rocks the name 'sial' (*si*lica-*al*umina) has been coined, while the rocks of the oceanic areas, typically dark, dense (density 3·3), are called 'sima' (*si*lica-*ma*gnesia). The oceanic rocks are to all intents and purposes thought of as basalts—formed from feldspars and dark minerals rich in iron and magnesium. It is true to say that nothing like the number of analyses of rocks from the continents will ever be available from the sub oceanic material. Much of the geology of the ocean floors has still to be directly observed and sampled. But in the oceanic islands, and among the titbits brought up by dredges and grabs from the depths, basalt seems to be the most common rock by far. Geophysical evidence, too, suggests that a relatively dense rock such as a basalt underlies the ocean floor. So there seems to be a fundamental geological difference between ocean floors and continents.

Although the continents stand high as elevations on the surface of the globe, the difference in elevation is small compared with the overall diameter of the planet. Earth is an almost perfect sphere, with a diameter of about 12,600 kilometres and flattened very slightly at the poles. The highest mountains may reach a height of some 8,000 metres; the deepest trenches in the ocean fall to 10,000 metres or more below sea level. From highest to lowest the relief is an insignificant amount on the curvature of the earth, and the high points and the abyssal trenches are but tiny parts of the global surface. Most of the surface lies at the ocean plains 3,000 metres below sea level with a step up to the continental platform at about 250 metres above sea level. A comparison sometimes used to illustrate the surface relief of the solid earth is that it could be represented *within* the thickness of a moderately thin pencil line drawn as a circle with a radius of four centimetres to represent the earth. If there were no geological cycle, no earth movement or vulcanicity, there would be no relief above sea level and even this pencil line would be much too thick to afford a useful analogy.

The Ocean Floor

Let us forget the water and look for a moment more closely at the surface of the earth in the great basins between the continents. Until very recently most of this domain has been unknown, but the wealth of bathymetric and geophysical data accumulated in the last two decades has given us a new view of much of the skin of the earth and how it has evolved.

We might best consider the relief of the oceanic part of the crust by making an imaginary journey from one continent to another across the ocean floor. Leaving the strandline where sands and boulders and muds may obscure the solid rock below, the descent is gentle out across the continental shelf. There are spreads of muds and sands right to the edge of the shelf and beyond. Most of this material seems to have been derived from the land. Currents, tides, storms and other movements of the water distribute it. Locally there are accumulations of shells, corals or other living or once-living materials. Life plays thus an appreciable part locally in producing high reefs and banks on the sea floor, but for the most part the relief is very gentle. At the end of our traverse across the shelf there may be a scarcely perceptible gradient developing away from the continent. In other places there is a sharp decline or even an almost cliff-like termination to the shelf. In a few places the continental edge is cut through by narrow, steep-sided canyon-like valleys debouching into the depths. They may be as much as 100 kilometres long, 10 kilometres from rim to rim and 1,000 metres deep. Many of them seem to be continuations of large valleys on the nearby land, and at the lower ends of some are delta-like spreads of sediment. One explanation for these submarine canyons is that during the periods of glaciation during the last million years sea level was lowered by the removal of water to form the ice caps and that large parts of the continental shelves were exposed to erosion by rivers. It is a neat hypothesis to explain many of the valleys but not all.

Skirting the edges of the submarine canyons with their curtains of unconsolidated sediment we should pass onto the continental slope. Now begins a descent of as much as 3,000 metres to the ocean plains below. In some places the gradient is steep and there is little sediment on the rock surfaces, but for the most part it is a gentle incline across sheets of mud and silt. It is not smooth and plane everywhere, locally there are swells and depressions, small valleys and ridges.

At around the 3,000 metre level the continental slope reaches the oceanic 'plains'—not that this is a transition that we are likely to notice. For many years it was thought that the ocean floor is indeed flat and plain-like, but its irregularities and complexities can now be

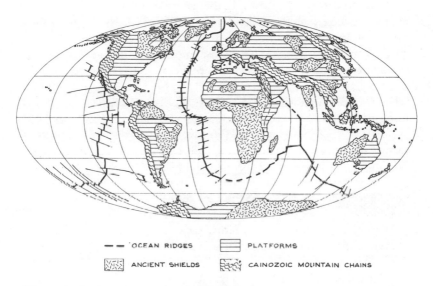

--- 'OCEAN RIDGES   PLATFORMS

ANCIENT SHIELDS   CAINOZOIC MOUNTAIN CHAINS

Figure 2-2 Ancient and modern parts of the crust exist side by side. The ocean ridges are being continuously renewed and the Cainozoic mountain chains are only a few million years old. The platforms are covered extensions of the ancient shields, the oldest rocks of which may be well over 3,500 million years old.

made out with some degree of precision in many regions. It is a strange 'landscape' with broad plains, plateaux and ridges and peaks, but lacking the valleys of the true continental landscapes. In many parts of the oceans there are clusters of conical peaks rising abruptly from the sea floor. The tips of many of them emerge as volcanic islands while others support coral roofs. Yet others stand as flat-topped submarine mountains or cones terminating well below the surface of the water. As far as we can determine, the coral-capped peaks and the sea mounts are all made of volcanic rocks. The flat-topped, steep-sided forms are known as 'guyots' in honour of a Swiss-American geologist, Arnold Guyot, who was a professor at Princeton University in the nineteenth century. The tops of these volcanic cones are thought to have been worn flat by the action of surface waves, since when either sea level has risen or the guyots have sunk. On balance, subsidence is thought to have played the major role.

We cannot travel far towards the centre of our ocean basin floor without finding that the ground beneath begins to rise gently, then very sharply for thousands of metres. Near the centre we approach what is in fact a remarkable ridge or palisade, not very wide but extending on each side as far as the ocean reaches. Some parts of these palisades are smooth, others very rugged and irregular, but all seem to belong to a network of ridges that divides up the deep oceans into smaller basins. It is not many years since the outline of this ridge system was first discovered, and in this time the system has become recognised as one of the most remarkable topographic features on the face of the earth. Here and there running along sections of the ridges are enormous escarpments, rather like those in parts of East Africa, and like them, they are probably the result of great fractures in the crust.

Were we to follow some of the ocean ridges our way would in places suddenly be cut off by transverse seas or scarps, to resume again but displaced to right or left by many kilometres. The Pacific Rise and the mid-Atlantic Ridge just north of the equator have suffered many such fractures.

Not far from the ridges in the western Pacific the ocean floor suddenly plunges into great arcuate or bow-shaped trenches. They are truly gigantic features, 200 kilometres or so across and 1,800 kilometres or more long. Since they swing around some of the great chains or arcs of islands in this region they are known as island arc deeps. The forces needed to produce such hollows are obviously not erosional. Where would the material from these deeps have been deposited? There is no sign of it on the local sea floor. At these cold, dark, great depths little moves, except perhaps the crust of the earth itself. Could it be that these furrows develop in response to the internal strains and stresses of the crust? Might there be a common agency responsible for producing the mid-oceanic ridge system, the sea mounts and deeps?

To these questions, many geologists would now answer 'yes'. Their reasons for doing so are marshalled in a later chapter, but for the moment we must pass on to the continents keeping in mind the outline of this oceanic natural architecture and its volcanic, basaltic monotony, together with one other important fact. The oceanic ridges and the deeps or trenches are extraordinarily prone to earthquakes. Compared with many parts of the crust, they shudder and shake with great frequency. The association of volcanic eruption and earthquakes is well known, and to our cost, in many lands. Under the oceans it is even more pronounced.

Once across the mid-oceanic ridge our traverse may take us across further oceanic plains, up the continental slope and on to a continental shelf again. En route there may be several kinds of sediment to examine layered over the basaltic rocks. Some of these sediments are made of fine clay, silt and dust washed or blown out from the land.

CONTINENTAL MARGIN | OCEAN BASIN FLOOR | MID OCEANIC RIDGE | OCEAN BASIN FLOOR | CONTINENTAL MARGIN

Figure 2-3 A distinguished profile is revealed in this representative section from New England to the north-west coast of Africa. It shows the steep drop from the continental shelf to the ocean basin floor and the tooth-like cones of the sea mounts and crests near the mid-oceanic ridge. The centre of the mid-ocean ridge is marked by a narrow rift valley.

Others are made up of the microscopic skeletons of tiny sea plants or creatures that have drifted down from above in countless millions. So slowly do these sediments or oozes accumulate that a thickness of a few millimetres may represent many thousands of years. These layers, too, have an importance in the story of the ocean basins.

So much for an imaginary traverse of an ocean basin, but the actual courses of the mid-ocean ridges across the globe should be mentioned. Starting in the Soviet Arctic near the islands of Severnaya Zemlya we could strike along a ridge across the Arctic Ocean basin, to pass north and west of Spitsbergen and south to Iceland. Southwards it runs along the curved mid line of the Atlantic past the Caribbean to pass between West Africa and Brazil. Here as at various other points on the way the ridge has been displaced by major transverse fractures or faults, each one south of 20 degrees north displacing the ridge to the east on its southern side. From 20 degrees south the ridge runs as a great loop some 2,000–3,000 kilometres off shore and around the southern cape of Africa and up into the Indian Ocean. A thousand kilometres or so east of Madagascar it joins a ridge which stretches to the coasts of Arabia in the north, linking with the great African Rift system of faults there. South-eastwards the ridge swings between Australia and Antarctica and eastwards into the South Pacific. There are small branches extending towards the eastern side of New Zealand and south into Antarctica and transverse faults cut the ridge at many points. A branch hits across towards southern Chile but northwards the ridge extends to the north-west coast of Mexico where it seems to find expression in the system of large faults in the Gulf of California and the south-western states of the U.S.A. It emerges again, offset from the coast of Oregon, to continue northward to Alaska. In the Pacific Ocean and the Indian Ocean there are many transverse faults offsetting the ridge by a few kilometres. Some of these faults run across the ocean floor for several thousand kilometres and they may displace parts of the floor in great escarpments. The Indian Ocean floor in particular is scarred by numerous north-south scarps and fractures of this kind. One striking feature of most of these faults is that they end at the continental rise and are not found on the shelves. Whatever it is that causes them only affects oceanic crust and not the continents.

The great deeps or trenches in the ocean floors are by no means as extensive as the ridges but they seem to form an almost unbroken chain from the Aleutian Islands, past Japan and the Philippine Islands, and from New Guinea to north of New Zealand. Other trenches run down the west coast of South America from Ecuador to Chile and on the west side of the Palmer Peninsula of Antarctica; a lone short trench lies in the South Atlantic and another east of the southern Caribbean Islands.

Again and again in this book we shall refer to the features that rim the Pacific Ocean and the earthquakes, eruptions and earth movements that take place there. We shall also see that the ocean ridges are the sites of a quite different kind of unrest in the crust, and they too experience eruptions and earthquakes. These wrinkles in the ocean floor seem to be some of the most fundamental features produced by the forces alive within the earth.

Volcanoes, Volcanoes

Whether we view it as it is, or in our mind's eye see the surface of the earth without its covering of seas and oceans, we cannot fail to notice that volcanoes, alive or extinct, form a conspicuous part of the relief. There seems never to have been a time in the history of the earth when volcanic activity was not going on in one place or another. On occasions, it must have been very much more vigorous than it is today, pouring out liquid rock or lava, coughing up solid dust and ashes and emitting great volumes of steam and gas.

Rather fewer than 500 volcanoes are active today or have been so in historic times; most of them are cones, a few are large flat domes. If the many extinct cores and craters that are still well preserved and geologically not very old are also taken into consideration the number rises into the thousands. The majority occur within well-marked bands across the globe, bands which are also marked by high incidence of earthquakes, by complex geological structures produced by earth movements and by high relief. Foremost among these regions are the great mountain belts around the Pacific Ocean—the Pacific 'Ring of Fire'—and the Alpine–Himalayan mountain ranges. The volcanoes in these regions have often been violent and explosive.

As we noted in the previous section, in the last two or three decades the mid-oceanic ridges and sea mounts have revealed an unexpected and overwhelming presence of basalt and volcanic activity that has quietly been in progress beneath the sea during the last 150 million years or so. The mid-oceanic ridges are, above all, sites of quiet but persistent volcanic activity.

Volcanoes are also present in other regions, notably Africa, where they occur close to the great faults of the rift valleys of East Africa and the Middle East and in the Saharan and West African territories. There is a clear association of earthquake activity and, as we shall see, faulting, in the African occurrences which extends back over the last 40 million years or more. The present spectacular volcanic peaks, Mount Kenya, Kilimanjaro and others are all comparatively recent in origin.

In West Africa, on the other hand, there is much less known about the relationship of the long chain of extinct volcanoes that extends from the Sahara through the Cameroons to the Gulf of Guinea. Perhaps these, too, are associated with fractures in the Great African Shield. It is certainly geologically unusual to find volcanic cores rising from what has been regarded as a rather rigid, stable area of Precambrian basement.

In the Mediterranean region a small group of volcanoes extends from Sicily and mainland Italy to eastern Turkey. But east of here the Himalayan mountain belt has a notable lack of volcanoes.

The Continents

To match an imaginary journey across an ocean basin by similarly traversing a continent may seem a superfluous exercise, but it may repay our patience. A quicker scan of the region of the earth above the waves is to be had from an orbiting satellite or other space ship and these means are being used to improve the quality and quantity of data on the more familiar parts of our planet.

Despite the crinkly, untidy coastlines, which are accidents anyway, the real outlines of the continents, drawn at the 100 metres below sea level mark, give fairly sharp, clear shapes. As many people in the past have suggested, they look like the scattered fragments of a single broken plate or pieces of a jigsaw. Their fascinating outlines need not detain us longer; more interesting for the moment are their internal anatomies. Strip away soils, alluvial deposits, the muds and gravels of glacial ages and recent coastal accumulations and the continents may be, like Caesar's Gaul, divided into three parts, or at least three kinds of terrain—shields, mountains and plainlands.

Of prime importance are the hard, granitic and crystalline areas of very ancient rocks. They underlie all other formations and are sometimes not inappropriately referred to as basements or shields. From these rocks have come the highest determinations of radiometric age, more than three thousand million years. They bear the scars of the great events in earth history since those times. Most of the rocks are metamorphic, changed by the heat and pressure of deep burial, and contorted by the years.

Some 89 million square kilometres of these shield rocks crop out today and are being closely scrutinized by the mineral-hungry nations of the world. Elsewhere the basement plunges beneath the great sedimentary ramparts of more recent mountain chains or lies not so far down beneath the great plainlands and continental interiors. Thus some areas of basement have from time to time over the last 600 million years or so sunk gently beneath the waves to emerge again unchanged except for their new thin coats of sediment. Others have foundered in crustal downfolds over a geological era or two, eventually to be pushed up again with much disruption and violence.

For the most part, however, we can consider the basement in each continent as a rigid affair, a slab in the earth's crust, warping here and there only if hidden forces are brought to bear. Where the force is of a particularly irresistible kind the slab may fracture rather than warp. Spectacular results are produced as witnessed by the magnificent parallel scarps of the great rift valleys of Africa and the Middle East.

Warp and fracture, the two ways of releasing stress in the crust, seem always to have affected the great basement slabs since the early aeons of earth history and the movement still continues. Earthquakes and volcanoes add to the interminable list of new things out of Africa.

For the most part, the basement rocks give rise to hilly rolling country; some regions are flatter, low lying as in Finland, others are mountainous as we see in Greenland and Baffin Island.

A closer look at the detailed structure of the old basement rocks reveals that they have many of the characteristics of the rocks found in the inner parts of mountain chains. Stepping off the basement rocks, we may in fact step onto those of a more recent mountain chain that is, so to speak, welded onto the edge of the basement, or thrust up onto

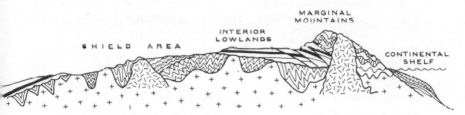

Figure 2–4 No two continents are alike but all are composed of rock formations which may be thrown into the kinds of geological structures and relationships shown here. Continents are patchworks of structures evolved over millions of years. To arrive at the state shown by this imaginary continent may have taken 3,000 million years.

it. Could it be that the basement has grown by the accretion of similar metamorphosed materials around a nucleus?

The mountain chains themselves are impressive enough in other ways. Composed of sedimentary, volcanic and local metamorphic rocks, some of them rise 8,000 to 9,000 metres above sea level. Long continuous ridges, rather than broad uplifts, they occupy conspicuous belts curving gracefully around the Pacific Ocean and from Spain and North Africa eastwards through the Mediterranean lands and the Middle East into the ranges of the Himalayas and southern China. In the Far East they link with the mountainous rim of the Pacific. Individual mountain chains, some decidedly curved, link together into bundles and groups. Within these areas may be volcanoes, extinct or active, and throughout the length and breadth of these highland areas there is always the threat of earthquakes. In many parts of the world the boundary between mountains and lowlands is very sharp. The foothills of the Rockies, the Italian Alps, the Grand Tetons, rise abruptly as daunting walls and cliffs of rock above the adjacent plains. Commonly there is a sharp break, a geological fault or fracture between the two. On the one side there is little disturbance, but the mountains invariably show signs of great compression and movement. Movement, volcanoes, earthquakes are here, as at sea, associated with the features of greatest relief on the crust.

The relief is produced not only by earth stresses pushing rock formations laterally and upwards but also by the extreme effects of climate and erosion. What mountain range does not have its deeply excavated valleys flushed out by streams or gouged by glaciers?

The annual monsoon rains, the snows and mists in their drenching of the mountains, carve and sweep out channels for themselves. The erosion is an incessant process which ultimately seems to have overcome most mountain chains, working, tearing and grinding them down to their very roots. No matter how violent and pervasive the internal forces were that pushed up the hills, gravity, the patter of raindrops, wind, the sun's heat and the crackle of ice finally reduce the land to less and less spectacular dimensions. Where mountains rise today seas once lapped, while in many lowland areas are the worn stumps and degraded remnants of once lofty uplands. One of the puzzles confronting geologists still is why some mountain belts seem to be rejuvenated from time to time while others crumble away quite rapidly. For one reason or another there seem to have been mountain ranges on the western side of the Americas throughout the last 600 million years or more. Spectacular as they are at present, their grandeur in one respect offers us a puzzle. At no time do the mountains of the Western Cordillera or elsewhere ever appear to have been much greater in height than the highest peaks today. There seems to be a natural limit to their height, beyond which they cannot rise.

Much of each continent is Precambrian, basement rock, a shield or nucleus. Equally distinctive geologically and geographically, the great mountains form no small part of most continents. But there are also wide parts of the continents that are neither shield nor mountain range, but are regions of perhaps monotonous geology where rock formations are flat-lying or cast only into gentle folds.

The land surface may be rolling and hilly or unbelievably flat, well-wooded lowlands or grassy plains or dusty desert, depending on its location. Its physical features probably reflect the outcrop pattern of the rocks, most of which are sediments, and some of which may contain fossil fuels, salt deposits and other useful raw materials. The need to prospect for such substances has brought in an enormous wealth of data about the rock formations of these areas.

Beneath the sedimentary formations ultimately there lies the Precambrian basement. Only locally is it near the surface or poking through the sedimentary cover. Elsewhere it is depressed into broad and sometimes deep basins filled with younger rocks. These buried extensions of the visible shields are far from being flat featureless slabs or pads but are warped into arches, domes, troughs and basins and may locally be fractured by faults large and small. Since Precambrian time, however, they have behaved as rather sober and slow-moving members of the crustal community. Periodically, extensive parts of them have been flooded by the shallow seas that have left behind the sedimentary cover. Coral seas and coal swamps have spread over some of these regions in their time, deserts and grasslands have appeared and faded in response to climate and the gentle movements of the crust. Small changes in land or sea level have more widespread effects where the relief is low and the land surface nears the strand level. These regions have been the settings for some of the most remarkable and widespread shallow seas of all time and they have had a profound effect upon the evolution and distribution of life.

Today's plainlands, low or high, grassy or forested, and even desert, continue to evolve. In the last million or so years those in the northern hemisphere have been extensively glaciated and are now being affected to no small extent by the activities of farmers, lumber jacks, miners and modern man at large. The seaward extensions of these regions are similarly in peril now for they constitute the wider parts of the continental shelves made accessible by technology.

Such Wear and Tear

With so much continental surface exposed to the depredations of erosion, geologists and others have been concerned to know how much material is removed from the land each year, and how much the rate has varied throughout time. Geomorphologists seek to account for the origin and evolution of different kinds of land forms. Their theories are based upon observations of the processes in action today and the shapes of the land surface itself. For them, erosion controls the shape of the land surface to an extent matched only by the geological structure of the ground itself. This is a good uniformitarian approach, but it is limited, as are many branches of geology, by the shortness of the time in which direct observations of the carving and building of the natural landscape have been possible.

Even so, this did not stop early geologists and geomorphologists formulating quite sophisticated hypotheses about the development of relief on newly formed land surfaces. They understood that once the raising of a new mountain by volcanic or tectonic activity or the elevation of part of the sea floor had begun, erosional processes would attack and reduce the surface in a sequence of events controlled by recognisable factors. Rock waste would be produced by weathering, transport would be provided by air, water, or ice, and what was left would reflect not only the nature of the underlying rocks but also the ways in which weathering and transport (and, locally, deposition) had gone on. If the land area was raised or lowered relative to sea level this too would be reflected in the rejuvenation of gradients and streams or by the deposition of sediments at sea level. Given no further uplift, however, erosion would ultimately reduce all the land to sea level. Interruptions in this process arise when uplift occurs and the business of erosion has to begin again cutting down the land to a new base level. It becomes a cyclic, or at least repetitive, process, occupying long periods of time. It produces sediment for new rock formations and land forms that themselves are either being buried under their own debris or whittled away to provide the sediment cover elsewhere.

The great exponent of this cyclic hypothesis of the evolution of land forms was the American, W. M. Davis. His work has influenced most of the study of land forms carried out in this century.

Davis estimated that the time required to reduce an upland region to a flat, plane surface, a peneplane, might vary between 20 and 200 million years. During the last forty years or so, however, new evidence points to a conclusion that present land forms may be much younger. The uplifted remains of old peneplanes form conspicuous level surfaces or common levels to the summits of hills and mountains, and deciding at what periods these planes were first formed and then uplifted is always a difficult business. In the 1950s it was believed that little of the present topography of the continents is older than the beginning of the Cainozoic, and most of it is no older than Pleistocene.

The various means of arriving at such a conclusion are concerned really with establishing the rate at which material is lost from the lands and swept into the sea. An obvious method is to find out how much material is carried in suspension and in solution by rivers. This is done by taking the simple but often rather arduous step of sampling

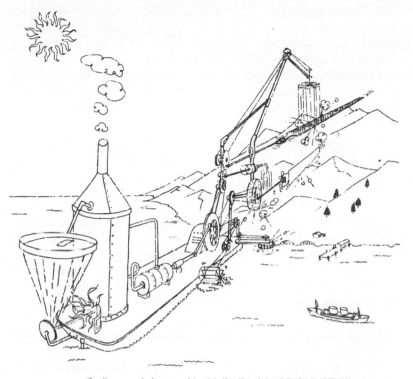

The "geomorphology machine" (after the style of Rube Goldberg).

Figure 2–5   The source of most of the energy expended on wearing down the land is from the sun.
(From Arthur L. Bloom, The Surface of the Earth, *1969. Reprinted by permission of Prentice-Hall, Inc.)*

the rivers. Rivers fluctuate during the year so the sampling has to cover this. In some parts of the world the behaviour of the rivers makes life hazardous and sampling can be a dangerous occupation. Most of the figures for rates of erosion of the continents have been based on estimates of this kind. Another useful measurement is that of the amount of sediment caught in reservoirs, but all sorts of corrections have to be applied for the effects of agriculture or other human activities. The measurement of such things as soil creep downhill, surface wash and landslides is helpful but obviously rather difficult to use on a big scale.

In recent years radiocarbon dating has been pressed into service, changes of land forms having taken place after an object datable by radiocarbon has been located. Buried wood, bone or other organic matter is essential here, and the radiocarbon isotopes only help for objects less than 40,000 years old.

But with such extremes of environment as deserts and rainforest, coastal swamps and mountain meadows, and with all the variations between them, it seems an almost hopeless task to gauge the erosion from the continents. Some generalisations can, however, be made. In the end, it might be said that there is little difference in the rates of erosion of igneous and metamorphic rocks, or of sandy and calcareous rocks. Unconsolidated sediments, however, are eroded very much faster than well hardened sedimentary rocks; 10–1,000 times faster according to some calculations. From the measurement of the rates and ways in which the land is worn away, two types of relief are clearly indicated. In the one category are mountains and regions of steep slopes, and in the other is what we might call normal relief, gentle slopes and plains.

A recent survey of erosion rates showed that it takes about 2,000 years to lower the land surface by 1 metre in mountainous regions but about 22,000 years to lower it 1 metre in areas of normal relief. To erode an upland region of continental proportions and 200 metres high to the condition of a peneplane would require at this rate about 10 million years. Of course lots of smaller areas have experienced greater rates of erosion in recent geological times either because of the Pleistocene glaciations or because of agriculture. Many regions, indeed, have experienced both.

So in 10 million years substantial changes can be brought about by erosion of the land. The continents all have a complex history of uplifts and erosion, the peneplanes produced at earlier times in the history of the land are all progressively worn away a little more with each phase of uplift. The older the peneplane—or the remnants of a peneplane—the higher it is in the topography. Some of the great plain surfaces of Africa, now several hundred metres above sea level, may well be mid-Mesozoic in age.

Geologists and geographers alike have long been puzzled by the way in which the shield areas seem to have had a tendency to rise time and again throughout geological history, keeping pace with erosion over the years. Suggestions that the continents were once smaller in area, but thicker, have been made and rejected. The simple calculation of the volume of sediments that might have been removed from the continents in recorded geological time gives a figure of 300 kilometres. Now such a thickness is impossible: it is more than ten times the thickness of the crust. We can retain this figure if we postulate either that the rates of erosion were very much faster in the past than they are today or that the material has been pushed round the geological cycle several times. As will become obvious on later pages, the second possibility has distinct attractions. There does seem to be a deep-seated cause to the continuing, if impersistent, role of the shields as regions of uplift throughout geological time and it is related to the behaviour of the crust as a whole. The shape of the land surface itself is ultimately related to the consequences of this in more ways than one.

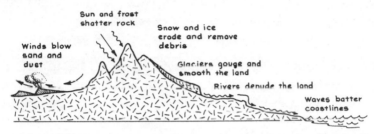

Figure 2-6   Some of the processes by which the land is weathered and eroded are easy to recognise. It is less easy to calculate exactly how much material they remove in a given period. Although their efforts may seem slow to us, they are achieved rapidly in terms of geological time.

The Mould on the Crust

Varied and interesting though the surface of the earth is in its own right, it is made even more so by the film of living things that is drawn over much of its surface. Only within the most arid or frozen deserts is it possible to stand and stare from horizon to horizon without sight of another living thing. Even at sea where the naked eye can rest on nothing but the waves, the lens beneath the water may reveal thousands of microscopic floating organisms, plankton. The deepest sea floors have their populations, thinly scattered but nevertheless significant for the biologist. In the jungles of Africa or Indonesia there may be little to see except living things above and around, and even the forest floor is littered with the remains of dead plants in process of biological degradation.

The conditions which control the distribution of life on land and in the sea are important in our story. If we can reasonably assume that the plants and animals were susceptible to the same kinds of physical influences in the past as they are today, we may be able to suggest where particular floras or kinds of animals could have existed in the past.

In happy instances there are abundant fossils to confirm our deductions. The process works in reverse too; finding certain fossils, we assume that the conditions their modern descendants enjoy prevailed when the fossils were alive. Understanding something of the ecological factors can help us picture more accurately how animal and plant communities may have flourished in the past. In the case of many extinct creatures it is difficult to work out the kind of environments that favoured them best; all that can be done is to apply the general rules that seem to prevail for modern animals.

At sea, life has adapted to several modes of existence. Floating organisms, *plankton,* and swimming organisms, *nekton,* are found in all but the very saltiest of seas and at most levels. Bottom-dwelling organisms, the *benthos,* on the other hand, may be attached (*sessile*) or able to move about as they wish (*vagrant*). Salinity and temperature affect the distribution of them all: some have wide tolerances, other plants or animals can survive only within narrow limits of saltiness and temperature. In addition to this, food supply, amount of daylight, currents, muddiness, and other factors all play a part. For animals there must be a supply of oxygen, and for plants carbon dioxide is needed.

On land, temperature and water supply are all-important. As rain, snow, dew or mist, the precipitation and humidity of a region may effectively control the range of plant life and this in turn is an important factor as far as animals are concerned. Temperature, the slope and relief of the land surface, the nature of the soil and bedrock also exercise prime controls on the spread of vegetation.

Broadly speaking, both land and sea floral and faunal communities are more varied in the tropics than near the poles. One can collect many more different kinds of shells, for example, on a sea shore near the equator than in the Arctic or Antarctic. Plants and animals on land are more numerous and various in the lower latitudes than in the high. Warm coastlines seem to have more marine plants and animals per square kilometre than the cold coastlines, though there are exceptions.

This matter is worth noting because it is in the shallow-water sedimentary deposits that there is the greatest chance of organisms —their shells or skeletons at least—becoming fossils. Indeed most of the record of past life is to be found in the sedimentary rocks that originated in the shallow sea. Freshwater and terrestrial organisms are less frequently found as fossils. The chances of non-marine plants and animals becoming fossilised are relatively slender. And the more complicated, delicate or fragile a shell or skeleton the less are its chances of becoming buried intact and sealed into the sediment to form a fossil when the sediment hardens to rock. For every fossil that is preserved there may be hundreds of other individuals of the same species that are totally destroyed, leaving no trace at all. In the case of vertebrates the ratio rises to thousands, and in the case of man it may be millions. Of course completely soft-bodied animals such as worms, or plants such as fungi, are fossilised only under the most extraordinary circumstances.

As in time geography changes, so do the environments in which fossilisation can take place. Drastic changes such as the onset of an ice age wreak havoc amongst the communities that have previously experienced only warm conditions. Volcanic eruptions and the diversion of warm or cold currents at sea can have immediate and catastrophic effects upon the local plant and animal communities. Earth movements and the changing patterns of sedimentation around the coastlines are continually affecting the areas occupiable by this species or that. But it is an ill wind that blows to no good effect, and as environments change there is almost invariably a newcomer ready to move into a niche abandoned by a previous species. When some organisms become extinct it may be for non-biological reasons or it may be that competition from other living things is too strong. The fate of all species is eventual extinction. The highly evolved and specialised are the most vulnerable; the lowly, unspecialised and tolerant have the strongest claims to tangible immortality.

We have no very accurate idea of the number of different kinds of species of animals and plants there are in the world today: it must be over a million. If we were to add all the kinds that lived in the past it would perhaps lie in the hundreds of millions. The highly evolved animals of the present, the vertebrates which include the fishes, amphibia, reptiles, mammals and birds, number perhaps only a twelfth or less of all animals alive today. About three-quarters of the total is made up of flies, moths, beetles and other insects. Yet how few insects are ever found as fossils!

Abundant though modern living things are, and numerous though fossils are, both as individuals and in numbers of species, they each occupy only a fleeting moment of geological time. A fossil, then, is a remarkable thing in that it is a petrified glimpse, incomplete and in black and white not living colour. Fossils, then, tell us about the evolution of life, about the way in which life has spread across the earth throughout its few hundred million years of existence, and about the kinds of communities and environments that have existed on the face of the earth during this time.

Living things are great destroyers of rock and they are also builders of rock, as we see from the limestone made by corals and shellfish. Life has played a vital part in the evolution of the earth and it is a characteristic of the planet today just as it has been for the last 3,000 million years or more. The biosphere—the mass of all living things on earth—is as much a part of earth as is the hydrosphere or the lithosphere.

Roots

Every aspect of the geology of the continents suggests movement, the movement of sediments from source area to site of deposition, the migration of the seashore and the heaving and pushing of the crust itself. Whatever else may be said about them, the continents have not been static. There must be a degree of plasticity in the rocks that allows them to be folded and heaped up into mountains. The question arises as to just what it is that permits the rise of mountains in well-defined belts such as the Alps or the Cordilleras of the Americas when the rest of the continental area is relatively flat.

As with so many important questions, it has taken a long time to find an acceptable answer. The first evidence to bear on the question has come from an unexpected quarter. In the mid-nineteenth century two explanations of what seemed to be an odd quirk of nature attracted the attention of geologists. Sir George Airy put forward to the Royal Society in London an idea to account for the failure of the plumb-bob to be deflected by the gravitational attraction of mountain masses by as much as in theory it should. This failure had first been found by Pierre Bouguer, a young French scientist and member of the Académie Français expedition to Peru (now Ecuador). The French expedition set out to measure an arc of meridian (as it did in Lapland) in order to determine whether the earth was really flattened towards the poles rather than at the equator, as some geographers thought at that time. Unfortunately Bouguer's account received little attention, but Airy advanced the idea that the crust of the earth floated on a denser substratum, like logs on water. The crust rests upon a plastic mantle, and the higher parts of the surface overlie thicker parts of the crust.

An alternative explanation was offered by Archdeacon Pratt, again to the Royal Society, in which mountains were thought to stand high because they are composed of lighter material. Then in 1889 C. E. Dutton suggested that the crust under the oceans was denser than that under the continents and he proposed the name 'isostasy' for the balancing of surface features by a substratum of differing density. He noted the singular fact that when prolonged erosion had removed a thick layer of rock from a mountainous area an uplift of almost the same magnitude took place. Similarly thick accumulations of strata seem to have sunk by an amount not much less than the thickness of the mass. The idea was hotly debated but with the advent of geophysical means of detecting subsurface structure it has become widely accepted. Mountains are thought of as having 'roots' projecting downwards into the sub-crust, the amount of root being proportional to the height of the mountain. The implications of this idea are far-reaching.

Water, Water Everywhere

Earth is truly the wet planet and, as we have remarked before, no doubt this is one of the reasons why the history of its surface has been so varied. Water is an important lubricant, it might be said, in the geological cycle. It can dissolve more substances than any other liquid known and it is essential for the support of organic life. In the history of our planet it has not only an important chemical function but also a significant physical attribute: it can store heat. The oceans give a fair imitation of a central heating system conveying heat to cool areas and circulating cool waters back to the hotter realms.

On a small scale this is easy enough to envisage but the oceans of the world contain over 1,000 million cubic kilometres of water. The circulation and behaviour of this mass are part and parcel of geological processes at the surface of the earth. Although temperature differences alone will cause currents, the flow of ocean currents is assisted by variations in the salinity of the water. Normal salinity is about 3·5 per cent and it has often been said that the oceans contain in total enough salt to cover the continents with a layer 100 metres thick.

The saltiness of the sea has probably not changed very much since the ocean basins were first filled about a thousand million years after earth first took shape. It has certainly been virtually constant over the last 200 million years, but the composition of more ancient sediments suggests that the ratio of sodium to potassium in sea water has risen from 1:1 to its present value of 28:1. Recently the idea has been developed that the salts in the sea are those remaining after a complex series of reactions has taken place amongst all the waters and sediments entering the oceans. It suggests that the weathering of igneous rock by rainwater containing dissolved carbon dioxide produces alkali and bicarbonate ions, clay minerals and silica. In the sea the clay minerals combine with the potassium ions. Marine organisms extract the calcium ions to make shells and other skeletal matter. Hydrochloric acid is added from undersea volcanoes and reacts with bicarbonate ions to produce carbon dioxide which may escape to the atmosphere. Meanwhile the clay minerals are incorporated in sediments to produce sedimentary rocks and the excess ions, sodium and chlorine, are left behind in the ocean.

There is no denying that these simple steps would have the effect of making the sea salty and something like this sequence does seem to take place. With a fairly constant source of chlorine from volcanic activity and a continuing removal of sediment from contact with the ocean by earth movements, this turnover could take place. Another aspect of this bit of geochemistry is that it suggests that the acidity of the oceans has not changed very much during the time this process has operated. And the implication of this is that the carbon dioxide content of the atmosphere has equally been held within narrow margins. On the basis of the inclusion of fossils, this seems certainly to have been the case for the last 200 million years.

The movement of the oceans has fascinated sailors, writers, scientists, painters. Tides, waves, ripples, storms, currents, have always pleased, frightened and amused mankind. Today's geologist needs to study them, for they are expressions of the energy that keeps the geological cycle moving.

The effects of such movements in shallow waters can be seen and locally measured; their contribution to eroding the land, transporting and depositing sediment and making things comfortable or otherwise for living things is easy to understand. From greater depths there is less information: it is clear that there are indeed slow but inexorable currents moving over the deep ocean floors. The complexities of deep ocean water movement are slowly being revealed. Many of the currents are in the opposite direction from those at the surface above them.

Tidal movements are well known in most of the waters of the world: the mariners' charts and pilots' handbooks give their range and the velocities of tidal currents. What is less obvious is that tides are experienced not only by the sea and all other bodies of water but also by the land and the atmosphere as well. A continental tide of as much as 20 centimetres has been calculated and an atmospheric tidal bulge of several kilometres is known. That tides are caused by the gravitational effect of the moon and to about half that extent by that of the sun, has long been common knowledge, but exactly how these effects are modified by the shapes and depths of the ocean basins is not so easy to understand. Giant tides prevail in some areas such as the Bay of Fundy in north-eastern North America and the Severn estuary in Britain. Here tides of over 10 metres are common. The amount of geological work done by the sloshing of water in and out of these basins is enormous.

Tides, storm surges and tsunamis (the giant waves produced by submarine earthquakes) are all capable of doing immense damage to coastal regions. They can, and do, produce catastrophes of truly biblical proportions and no doubt many such an event in the distant past has left its trace in the geological record.

With the boom in oceanographic studies in the last 30 years or so has come the realisation that not only does the water in the oceans circulate but that there is indeed a sort of ocean current 'weather'. There is a broad pattern to the movement of ocean currents and there is also a remarkable degree of variability, in the same way that there is not only climate but weather. The ocean circulation, like the atmosphere, is fickle, full of vagaries, background 'noise' and unpredictability. Even as well-known and trusted a current as the Gulf Stream, flowing north-eastwards across the North Atlantic, can change its strength and its position by scores of kilometres in a few hours. Ancient navigators may have trusted other currents to carry them to their destinations, but as Thor Heyerdahl on his *Kon Tiki* and *Ra* has shown us, such currents offer an uncertain guarantee of passage.

Before World War II the extent of variability in the strength and paths of ocean currents was virtually unknown. Since then improved navigational aids and new technology have helped oceanographers map the general circulation of the world's oceans rather as the meteorologists determine air currents and other parameters at specific places. Measurements of salinity, temperature, atmospheric pressure and other changes back up the measurements of water movement.

What appears from these studies is that the ocean waters circulate in the true sense; they move in circles or gyres, corresponding largely to the wind belts of the atmosphere and rotating counter-clockwise in polar regions and clockwise in sub-equatorial regions. There is a narrow gyre on each side of the equator and a counter-clockwise gyre in the sub-tropical belt above and below the equator.

Each gyre has a strong persistent current on its western side in response to the eastward rotation of the earth. Compensating currents operate in the central and eastern parts. This is part of the Coriolis effect whereby trade winds blowing towards the equator from the temperate zones are deflected to the right and blow from the north-east in the northern hemisphere and in the southern hemisphere are deflected to the left and blow from the south-east. Since we have not one ocean but several, the pattern is modified in each according to the shape and size of the ocean.

In modern oceans the gyres correspond closely not only to wind systems but also to chemical and biological patterns. At the boundaries of each gyre the conditions may change sharply. Cod fishers off Newfoundland or Bear Island will know where the cold and warm currents are from the way in which the fish crowd into the cold streams leaving the warmer water deserted above or below. At the centres of the sub-tropical gyres is a relatively warm stable mass of water, still and salty, poor in phosphates and biological activity. The Sargasso Sea is one such centre, a doldrums area feared by the masters of sailing ships.

Our interest in ocean currents is keen because they are intimately related to world climate, wind circulation and the dispersal of solar heat. They help or hinder the migration and distribution of life in various ways today, and presumably did so in the past.

Knowing something of the way in which modern ocean currents behave, we may be able to suggest how the currents in the ancient oceans moved and this should help us in reconstructing the climatic patterns of those bygone days.

Climate

Climate is part of the environment. It governs the characteristics of the flora and influences the fauna. Some writers hold that it produces the dour mien of Scotsmen, the sunny dispositions of the Mediterranean people and the adaptability of the English. At the root of it all is the effect of the sun at different latitudes and of the inclination of the earth.

At school we learned about the overhead sun of the equatorial midday and the six-month night at the poles. The effects of the angle of the sun's rays and the passage of the earth around the sun bring summer and winter as regularly as day and night. On these events hinge precipitation, atmospheric pressure, wind directions and speed, cloudiness and other attributes of the weather. Always we search for patterns of cause and effect because most people take an interest in the weather not so much for what it was yesterday, or is today, but for what it will be tomorrow or when we wish to set sail, go fishing, or climb a mountain.

Climate has seasons or cycles which to some extent we can relate to those of astronomy—24 hours, $365\frac{1}{4}$ days, and so on. The regularity of these weather cycles is not to be trusted, in the short term at least, as every Briton knows. In many regions of the world one may generalise about the climate but the weather may not conform to rule.

There is abundant evidence that climates change. Climatic fluctuations are known from the geological record as well as from history. All are puzzling. Some fluctuations are small, a few decades; others are of centuries and millennia. The reasons for them are hidden, but we can trace certain trends. For example, in the first half of this century there was a gradual warming of the globe, felt more especially in the Arctic than in the temperate areas and not experienced in the southern hemisphere. By 1960 it seemed that this phase had come to an end. Similar fluctuations occupy recorded history and before the advent of civilisation there were the last great ice ages.

On previous occasions as well as in the Pleistocene epoch there have been glacial phases when the overall temperature of the earth must have dropped and snow and ice were very widespread for thousands, perhaps millions of years. There was at least one, and possibly more, before the Cambrian period dawned; Permo-Carboniferous rocks in the southern continents record a glacial period and there may have been such a lowering of temperature locally in the Ordovician period. Each of these glacial events was a long complex shifting of climatic effects, not a simple cold snap.

What produced these ice ages is a question hard to answer. Various theories have been contrived, calling upon extra-terrestrial influences, geological changes or both. None is completely satisfactory.

The concept of continental drift can be used to explain many geological phenomena, but the implications of the regrouping and migration of continents for world climate have not been fully debated. At least two geologists have recently gone on record as saying that the idea of continental drift for the southern continents is incompatible with the evidence of coal swamps and Permo-Carboniferous glaciations there. The evidence of past climates is there, our interpretations may be inaccurate, but as our notions of the Permo-Carboniferous geography clarify we may be able to resolve such problems.

For the greater part of earth's geological history the world climate seems to have been rather warm and certainly more uniform than it is now. Climatic zones, as we know them now, seem to have developed only during the Cainozoic era. When glacial climates were produced they were short-lived in comparison with the warm periods. It is difficult to see where the common factors for the ice ages may lie, but this will not prevent our speculating about them.

CHAPTER III

Darkest Earth

WITH such an enormously varied planet to study, it seems a little hard on earth scientists that their opportunities for examining what happens beneath the surface are so limited. The oceans, atmosphere and the upper layers of the crust or lithosphere are now fairly accessible but below the limits of the deepest mines and boreholes, only indirect evidence of the nature of our planet is available. There is not much doubt that the evolution of the earth and of the peculiar nature and distribution of its surface features has very much depended upon what has happened at greater depths. And if the exterior of the earth has evolved with time is it not possible that the nature of the interior has changed too? The idea that it has changed has found favour with many geophysicists in recent years. Their reasoning is ingenious enough and their conclusions may be correct, but most geologists prefer as simple a model (or idea) of the interior of the earth as possible and hesitate to imagine conditions in the deep interior changing much from one age to another.

Two kinds of fairly frequent events, however, do present a means of finding out more about the nature of the deeper regions. Volcanic eruptions and earthquakes are among the most impressive phenomena that occur on the earth. No one who has lived through either is ever likely to forget it. Both are outward signs of equally impressive and important processes at work below. The two may occur together. Many a volcanic eruption is heralded by sizeable earthquakes, and while eruptions are taking place there may be a succession of minor tremors and quakes. The most violent and destructive earthquakes may occur in regions where there is in fact no active volcano at the time, but these regions have in the past, or may in the future, witness violent volcanic outbursts.

The Fiery Underworld

Volcanoes are, as we have seen, holes in the earth's surface through which heat and matter escape from within. They have fascinated men since the days of prehistory and even today their activities are difficult to account for and hard to predict. No two volcanoes are alike. They change their shape and behaviour from year to year and they may arise, die and decline with uncanny speed. Little wonder that volcanoes, with their fires, explosions, clouds of ash and billowing vapours, have been regarded as gods, angry spirits, or the gateways to the inferno. Aristotle believed volcanic activity and earthquakes to be related. So keen was the Roman naturalist, Pliny the Elder, to witness the eruption of Vesuvius in AD 79 that he stayed too long near the erupting vent and was overcome by the fumes. Perhaps he was the first geological martyr.

Despite centuries of observations, not all of them very reliable of course, the volcano enthusiasts—they would prefer to be called vulcanologists—are only just beginning to understand what leads to a volcanic eruption and what happens during the course of it. It is an uncomfortable job, often involving being perched on the rim of a volcanic crater or prodding the glowing ash or lava. Seismic observations help to reveal the state of things below the surface, rather as a doctor uses a stethoscope to listen for movements within a body. Tiltmeters on the ground surface may reveal tiny changes in the topography around a volcano approaching, during and after eruption.

There are basically two kinds of volcanic eruption. 'Quiet' eruptions occur when lava issues from a fissure or vent in the ground, making a good deal of noise and smoke or steam as it does so, despite the label. 'Explosive' eruptions, on the other hand, are just as the name indicates. Some explosions are relatively mild, others very violent and large, far exceeding the largest nuclear explosions. It is these explosive volcanoes that build up the well-known conical shape with the central vent at the top. The core is the ejected pile of debris, solid ash and molten lava that comes to rest near to the seat of the outburst. Eruptions may be closely spaced or sporadic over the years; the volcanoes may be dormant for hundreds, even thousands, of years. During this time gases or water vapour may be accumulating below the surface until there is sufficient pressure to blow out the congealed lava or ash that obstructs the exit. When that point is reached the result may be that the volcano literally blows itself to pieces.

The actual lava emitted from these various kinds of volcano is by no means a simple and uniform kind of rock material. Most of it approximates in composition to basalt, that heavy dark rock composed of feldspars and iron and magnesium-rich silicates.

Most of the volcanoes associated with the oceanic ridges have primarily basaltic lava and ash. On the other hand, much of the lava and ash from the explosive volcanoes which are found at the edges of the continents around the Pacific and also in the Mediterranean belongs to the type of igneous rock known as andesite. Andesite contains rather more silica than does basalt, and gives a less freely mobile lava.

Volcanic eruptions are given their explosive power by the presence of an unexpected ingredient, water. Water seems to be an essential part of molten rock, or *magma*, and as *magma* nears the surface the water tends to separate from the rest and push upwards like bubbles in soda water. At the high temperature beneath a volcano this water is

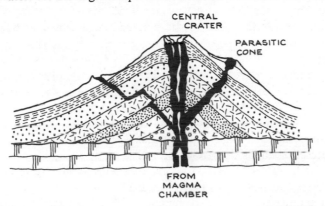

Figure 3–1   A volcano may take the form of a steep cone such as this, built up by layers of lava and ash over the years. Periodic explosions may shatter the cone and parasitic cones may mask the ideal conical shape.

present as water vapour and it exerts a tremendous pressure upon the plugged volcanic vent. Once the pressure exceeds the strength of the cover an explosion occurs. The escaping steam and gas expand madly, shattering rock and scattering *magma* into the air.

Several tremendous volcanic explosions have occurred in historical times, causing great loss of life. Fortunately they are not common, but the catastrophic eruption of the island volcano Krakatoa in Indonesia in 1883 was responsible for the death of over 36,000 people. The several explosions were heard as far away as India and Australia, 5,000 kilometres (3,000 miles) away! The fate of the Minoan civilisation of Crete may have been sealed between 1400 and 1300 BC by the even larger explosion of Santorini not far off in the Mediterranean. Both of these eruptions may have removed volcanic cones that rose over 700 metres above the sea, leaving in their places deep jagged craters on the sea floor. Explosions on a smaller scale may happen from time to time as at Vesuvius and Mount Etna.

Happily for mankind most volcanic eruptions are not like this. They may involve the movement of no less material but they do it in a more gentle fashion. Many of the volcanoes around the margins of the Pacific Ocean have emitted dust and lava as well as gas for a long time, building up beautiful cones around the vent.

The tidal waves set off by volcanic upheavals can be immense; in the case of Krakatoa the tidal waves crashed onto or swept over surrounding lands to destroy towns and villages in the space of a few moments, and in fact most of those killed by the events at Krakatoa were drowned.

Other volcanoes send out vast quantities of steam, gas and ash for several months or years and then die away. Paricutin in Mexico is such an example; erupting for about nine years, it produced a mountain where previously there had been flat fields.

Most fearsome of all eruptions are those that involve sudden explosive eruptions of gas-liquid or gas and dust mixtures. They eject incandescent clouds that move at speeds of up to 80 kilometres per hour and travel as far as 100 kilometres. With temperatures of many hundreds of degrees these clouds can do immense damage.

Yet another type of volcano and eruption is typified by Hawaii and the volcanoes of Iceland. The lavas break out fairly frequently but do so less explosively. Gas and dust may shoot out for a while, but lava in great quantities is the especially distinctive feature of these volcanoes. It forms low domes rather than steep cones as it escapes from the vents. Being comparatively fluid and easy moving, it flows away rapidly from the central crater or fissure from which the eruption is taking place. Often at the centre of these volcanoes the lava sloshes and bubbles in a lava lake within the crater, rising only occasionally to flood or dribble over the sill and down onto the surface roundabout.

These volcanoes are sometimes called shield volcanoes because they build up wide, thick shields or mounds by piling flow upon flow. Most of the material is dark basalt, which may be comparatively fluid when molten, spreading out quickly in flows and rivulets from the rent. The largest shield volcano—or for that matter, any kind—is certainly Mauna Loa in the island of Hawaii, reaching from the ocean floor 5,000 metres below the surface to over 3,000 metres above, and some 175 kilometres across its base. It has taken more than a million years to reach these dimensions and it may be one of the oldest of the volcanoes still active.

.There is yet one more kind of eruption to mention. It involves the outpouring of immense quantities of basaltic fluid lava which come from great fissures or eruptive centres set in lines. These fissure eruptions may build up hundreds of metres of lava flows covering thousands of square kilometres. Instances of them occur in Iceland with fragments of old flows preserved in Greenland, Scotland, Ireland and the Faroes. Others are known covering parts of the states of Washington, Oregon, Idaho and California in the U.S.A., the Deccan plateau of India, Ethiopia, the Zambezi-Victoria Falls areas of Africa, and also in South America and Australia. The Icelandic sheets of basalt may have covered more than 480,000 square kilometres. It would be wrong to view volcanoes, however, as taps to the entire inner reservoir of liquid rock within the earth. In most volcanoes the molten material seems to come from a limited reservoir not very far down. Once the volcano has emitted the contents of this reservoir, it lies quiescent until a new quantity of hot liquid rock or magma has been generated.

The presence of volcanoes and volcanic activity depends upon the presence of two basic phenomena. One is some form of fracture or other plane of weakness up which magma can force its way, and the other is a plentiful supply of magma at depth. All, or nearly all, our evidence suggests that the two are inextricably linked. Whatever it is that activates the crust or the mantle beneath it into producing magma also seems to cause earth movements not only associated with actual eruption but longer lasting and more extensively spread. One might imagine that magma could rise only where tensional cracks existed, but in fact the Pacific 'ring of fire' is one of the regions of the world most abundantly displaying signs of great compression. Tension or compression, both are forces capable of producing fractures in the hard crust of our planet. If there are areas that clearly are being compressed, might there not be elsewhere regions where tension prevails, as a sort of compensation or reaction? The idea has flickered in the minds of geologists now and again, and we shall return to it on a later page.

Earthquakes
Perched on the sides of volcanoes are volcanic observatories. In most cases the surroundings are not very hospitable and from time to time

an observatory has to be evacuated or removed; some are destroyed by volcanic action. Volcanoes are commonly rather local, if violent, in their activity. Earthquake observatories, on the other hand, have seldom to be evacuated because of the danger of earthquakes.

Nevertheless, earthquakes are widespread and can be violent enough to constitute one of the major geological hazards man has to suffer. There are still rather few earthquake, or seismic, observatories, and most of them are located either in the technologically advanced countries or in the major earthquake belts—the lands bordering the Pacific Ocean and the Caribbean and the mountainous Alpine–Himalayan-Indonesian belt. Some of these regions are famous—or infamous—for the violence and number of earth tremors. Japan, for example, lies in a belt where earthquake activity is at its strongest. On average, there are six major shocks a year there, with two or three minor tremors to be detected every day. In California major shocks are thought to occur about every sixty years with many very small quakes in between. Earthquake engineering, the design of buildings and other constructions to withstand the effects of the violent disturbances, has become normal practice in these countries. Elsewhere, many smaller quakes have little effect but the advent of an unheralded violent earthquake at rare intervals may cause tremendous damage and loss of life.

Although an earthquake is a momentary disturbance, its effects may be no less spectacular than those of a volcanic eruption. In addition to the shock experienced by the rocks below the surface there may be landslides and rock falls triggered off by the quake and around coastlines seismic tidal waves may wreak immense havoc. In towns and cities the actual destruction produced by the tremors in the ground may be enlarged by fires. In some Japanese earthquakes relatively little damage was done in the cities early in this century but the ensuing fires that sprang up virtually destroyed all that was left. Water mains were fractured and useless. Thousands of casualties occurred amongst the townsfolk.

The quake itself may range from a scarcely perceptible thump or shudder in the ground to a violent shaking or vibration spread over several seconds and recurring over several minutes. Strange growling and rumbling noises and sharp reports may be heard from below ground. Cracks and fissures can appear in the ground while landslides and rock falls may take place for days after the main shock. Streams and lakes may be disturbed, and the muds on lake or sea floor can be thrown up in heaving, stinking waves.

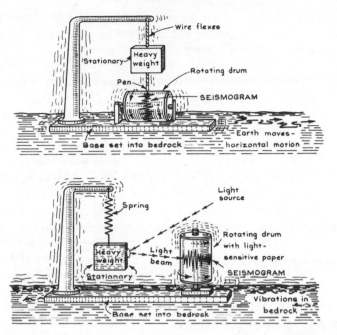

Figure 3-2 Seismography instruments such as these work on the principle that if the earth quakes the instrument is shaken but the suspended heavy weight remains stationary. Horizontal movement is recorded by the upper device and vertical movements by the lower, but in fact seismographs are refined to a degree far beyond the primitive arrangements shown here. (After Fagan.)

In the Pacific Ocean giant seismic tidal waves radiate out from where an earthquake is centred; travelling at several hundred kilometres an hour they may lower or raise sea level by 10–20 metres or more. The damage they can cause to shipping, installations and life is almost unimaginable. The Hawaiian Islands have been hit by thirty or more such waves since they were discovered by Captain Cook in 1778.

Within the last two decades an international network of seismic observations together with oceanographic research has established that the belts of earthquake activity on the continents connect with some of the great ridges on the ocean floor. Earthquake activity has much the same world distribution as have volcanic outbursts, and in the case of the Atlantic and other oceanic ridges more than twenty earthquake swarms have been detected in the last ten years, associated with faulting, magmatic and other processes. Earthquake swarms are sequences of quakes closely grouped in space and occurring within a short span of time; but there is no one outstanding shock or tremor.

At this point it would be well to be more specific about the nature and origin of earthquakes. To assume that earthquakes produce volcanic activity would not be correct, nor would the statement that volcanic activity is responsible for earthquakes reveal the whole truth. The explanation for some earthquakes is that the heat so typical of volcanoes is continually being produced deep within the earth and needs constantly to escape. Between those regions of the crust where volcanic or terrestrial heat is strong and adjacent cooler rock strains develop and are relieved by the breaking and movement of the rock. Many earthquakes, however, appear to originate along existing planes separating one slab or block of the crust from another. These fault zones, as we may call them, are active because the slabs or blocks themselves seem compelled to move relative to one another from time to time. Over the years that the San Andreas fault in California has been known it has been recognised that the land on the western side has been nudging its way northward in a series of rather regularly well-spaced jolts or quakes. San Francisco unfortunately seems to be sited right on the fault line.

To some extent the rock can yield to these strains by bending, or by plastic flow. The adjustment is continual and non-violent. A stronger strain may exceed the plastic limit of the rock and it has to respond by fracture. A likely analogy is that of bending a stick of candy until suddenly it breaks with a loud snap.

The snap is the result of the sudden release of energy by the break. From the site of the actual break the sound waves carry outwards in all directions and from distant surfaces there may be echoes of the report. When an earthquake occurs there may be an actual snapping of the rock, releasing tremendous energy, or there may be an equally sudden release by movement along an already existent fracture. Even a readjustment of only a few inches along such a plane will involve millions of tons of rock. The energy released in any one of the ten or so major earthquakes that occur each year is about 1,000 times as much as in the atomic bomb set off at Bikini Atoll in 1946. Twenty kilotons of T.N.T. is the estimated strength of that explosion.

The shock waves produced by earthquakes travel round and through the earth. In recognising the paths they follow and the speeds at which they move the seismologist may reach useful conclusions about the kinds of media through which the waves have passed.

Together with the results of laboratory experiments on rocks at high temperatures and pressures, and some theoretical physics, they can inform us about the conditions within the earth. Scientific seismology involving instruments suitable for world-wide use began late in the nineteenth century when John Milne, an English professor of geology in Tokyo, set up an observatory there.

It was another English seismologist, R. D. Oldham, who in 1897 recognised the three main kinds of seismic waves recorded by the instruments. They are primary waves (P waves) of the push-pull kind, like sound waves, secondary waves (S waves) in which the vibrations are at right angles to the direction in which the shock wave is moving, and longitudinal waves which move over the earth's surface.

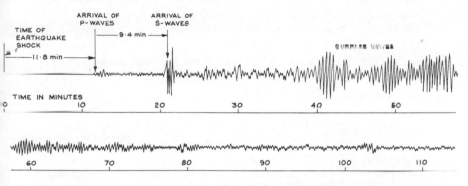

Figure 3–3   Autographs of an earthquake. This seismogram records a 'quake that had its epicentre over 8,400 kilometres away (around the surface). The separate identities of the P, S, and L waves is quite clear. (After L. Don Leet, 1950.)

The exact point at which the earthquake is generated within the earth is known as the *focus*, and the *epicentre* lies at the surface immediately above. Generally, the greatest movement and damage is felt at the epicentre, diminishing in all directions away from it. By simple calculations the position of a focus can be fixed where three or more seismological stations record the shocks sent out from it. The P waves can travel through both solid and fluid parts of the earth but the S waves move only through solid where they reach about two-thirds of the speed of the P waves. It is known that the speed of the P waves increases with depth in the earth and can reach almost 14 kilometres per second, the speed being controlled by the density of the rock.

The routes followed by the shock wave are not straight but, because of the changing properties within the earth, they have rather curved paths, being bent and reflected as they pass from one layer to another. At the surface they are reflected downwards again. In 1906 Oldham described how although big earthquakes send waves out in every direction, some parts of the surface seem to escape any movement. These parts lie in a sort of shadow zone in which P waves are subdued or suppressed. Oldham used this evidence to show that the earth has a large central core. The core throws a 'shadow' onto the side of the earth opposite a big quake. Not long afterwards, in 1914, the German geophysicist Beno Gutenberg located the top of this core at 2,880 kilometres (1,800 miles) below the surface, which means that the core has a radius of about 3,250 kilometres (2,200 miles). To this day the top of the core is called the Gutenberg discontinuity which is a fitting name for it in view of Dr Gutenberg's very careful work.

Gutenberg's discovery of the size of the core was a major event in seismology. The remaining part of the earth outside the core is known as the 'mantle'. The next thirty or forty years in seismology was a period in which Gutenberg and others most painstakingly compiled a series of tables for the times of travel of P and S waves along the various sections of their routes. They had of course to await the occurrence of earthquakes and to have some very delicate and accurate instruments ready to record the shocks. It was found that S waves do not travel through the core. On this basis and with some other evidence the fluid character of the core seemed established. A Japanese researcher calculated that this material could only be one-300th as rigid as the material just above.

To use the words 'solid' and 'fluid' for materials under such huge pressures as prevail within the earth may seem odd, but they refer to the elastic properties of the stuff. The solid or rigid material resists pressure, a shear stress. A fluid, on the other hand, has almost no resistance to shear stress and hence does not transmit S waves. The mantle of course transmits them very well.

There is, however, another layer to be distinguished above the mantle, the crust, and the distinction was first made by a Croatian seismologist, A. Mohorovičić. He was studying a Balkan earthquake in 1909 and found an important new characteristic in the way in which waves passed through the mantle. At a rather shallow depth some waves are reflected back to the surface and there is a sudden increase downwards in the speed of the P and S waves. Clearly there is a sharp change in the nature of the material at this level. It marks the boundary between the mantle below and the crust above. The boundary is known as the Mohorovičić discontinuity, or simply the Moho, lying at about 32 kilometres (20 miles) below the surface. One says 'about' 32 kilometres, because it is clearly very much less under the oceans and may locally be much more beneath the continents. The crust is less rigid than the deep mantle and is penetrated by many irregularities.

The seismic reverberations from within the earth, and especially those that tell us specifically about the crust, emphasise the point that crust makes up only about 0·4 per cent of the mass of the earth. Core and mantle, making up 99·6 per cent, are beyond our view. Even the largest features in the crust are seen to be infinitely small in comparison to the total size of the planet. It seems almost an impertinence to talk about the history of the earth when we can only acquaint ourselves with so little of its bulk.

The core, too, has been studied further from seismic records and is now known to consist of an inner solid part, about 1,220 kilometres (760 miles) in diameter surrounded by a shell of 'liquid' material, another 2,200 kilometres (1,375 miles) thick.

With experimental evidence about the behaviour of rock materials under great pressure and at very high temperatures, and using the fact that the speed of earthquake waves reflects the density of the rocks passed through, several scientists have estimated that the earth's density increases gradually from 3·3 grams per cubic centimetre at the top of the mantle to about 5·5 at the bottom and then suddenly reaches 9·5 grams per c.c. at the top of the core. At the bottom of the outer core it may be as much as 11·5 grams per c.c., and between 12·5 and 18 at the centre, a great density indeed but not unexpected when one recalls that the density of the earth as a whole is 5·52. We find it difficult to imagine how matter of such density behaves, but it was recently suggested that because there seems to be a tendency for the inner core to shear easily it might be compared in behaviour to fudge. One might say that the earth was thus soft-centred, with a core mostly of solid grains but with a very little liquid between them. Even in material like this there could be some convective movement of the liquid, at least enough to keep the magnetic field active.

K. E. Bullen, an Australian geophysicist, gave some interesting figures to help conjure up the magnitude of the pressures deep in the earth. The weight of the atmosphere at sea level is about 1 bar, at the bottom of the Pacific Ocean the pressure is equal to 800 bars. At a depth of only 320 kilometres within the mantle the pressure is 100,000 bars, which is about as high a pressure as can be reached artificially in the laboratory. Where the mantle joins the core the pressure could be 1·75 million bars and at the centre nearly 4 million bars. Such is the pull of gravity.

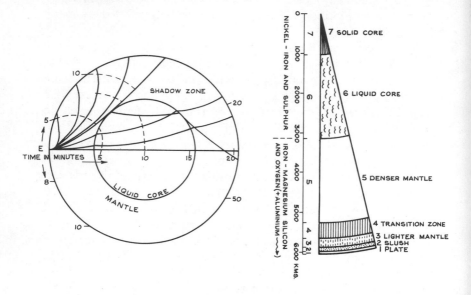

Figure 3–4   *Left*. Earthquake shock waves follow curved paths through the earth, their direction and speed being determined by the nature of the material through which they pass. *Right*. The interpretation made from very large numbers of earthquake records suggests seven zones within the earth of which the outer is relatively stiff. Zone Two seems to have fluid properties in part and could be a region where slow movement is in progress.

Calculations tell us that at the base of the mantle the material reaches the rigidity of about four times that of steel at the surface. Then at the level of the inner core the rigidity is suddenly lost, meaning that the dense material there is nevertheless fluid in some respects.

These calculations are all very well in their way, but do they indicate any kind of material we can understand? What is the deep interior of the earth made of? Ideas on this have changed over the

Pl. 1. Messier 81, a spiral nebula in the constellation Ursa Major, photographed through a 200-inch telescope. In its early stages the solar system may on a very much smaller scale have resembled this. *Courtesy of the Mount Wilson and Palomar Observatories*

Pl. 2.   *Above* Spiral nebula in Virgo seen edge-on. Messier 104 seen on 200-inch Hale telescope. The way in which most of the nebula is gathered into one plane is well shown here.   *Courtesy of the Hale Observatories*

Pl. 3.   *Opposite* A gaseous nebula in the constellation Gemini, photographed through a red filter on a 48-inch telescope. Such nebulae, the products of exploding stars, give off unimaginable amounts of radiation. An exploding star nearby could have drenched the surface of the earth in radiation, causing the extinction of many organisms.   *Courtesy of the Hale Observatories*

Pl. 4. The Earth seen from Apollo 8. The western hemisphere from eastern Canada to Tierra del Ferego and the 'bulge' of West Africa are visible where not obscured by white clouds. *Courtesy of National Aeronautics and Space Administration.*

Pl. 5.   At Siccar Point in Berwickshire, James Hutton saw that the Old Red Sandstone overlying the deformed and eroded Silurian rocks contained pebbles of the Silurian material and he began to unravel the evidence of the long history of the earth's crust here.   *Crown Copyright Reserved. Courtesy of the Institute of Geological Sciences*

Pl. 6.   This highly altered and banded rock known as gneiss, from north-west Scotland, was produced deep in the crust by the action of heat and pressure late in Precambrian times. Together with granite, it is one of the commonest kinds of rock in the shield areas of the world.   *Crown Copyright Reserved. Courtesy of Institute of Geological Sciences*

Pl. 7.   An aerial photograph of part of the Canadian Shield. The different rocks are covered only locally by soil or vegetation and so their distribution is clearly seen. Here gneisses are intruded by a mass of granite and granite veins. A major old fault may be indicated by the narrow zone that runs diagonally from the bottom left-hand corner.   *Courtesy of the Geological Survey of Canada*

Pl. 8. The Volcano Cerro Negro in Nicaragua erupting in 1968 rapidly builds up a core of ash and lava. Small parasitic cones let off steam, gas and ash from time to time. Volcanoes of this kind are typical of Central and Andean South America and parts of the west coast of North America. *Courtesy of Mark Hurd Aerial Surveys*

years as knowledge of rocks and minerals has grown and as the chemistry and physics of other planets and of stars has become known. For many years it was thought that the mantle below the crust must consist of rocks composed largely of magnesium-iron silicates. However, we now know that at the top of the mantle minerals like olivine and pyroxene, which are the common dark constituents in basalts, are abundant. Deeper into the mantle, garnet becomes important and other denser minerals occur such as diamonds, which cannot be formed except under enormous pressure. Diamonds are not only a girl's best friend, they are a delight to seismologists because their presence in rock shows that it has come from regions of extreme pressure. The pressures commonly found in the crust allow carbon to exist as a black very soft mineral, graphite, but at the great pressures and temperatures within the mantle its atomic structure is shaped in a cubic pattern and produces the hardest minerals known. Synthesis of diamonds has only been accomplished in equipment that can simulate these pressures and temperature conditions. Most writers have suggested iron sulphides, and nickel-iron is a popular notion for the core. Some meteorites have this composition, and they are thought perhaps to be fragments of a shattered planet once like the earth.

Whatever its composition, the material at depth, and hence at great pressure, is likely also to be at a high temperature. Unfortunately there are far fewer ways yet known of finding out about the internal heat of the planet than about its pressure. Within the range of the deepest mines the temperatures increase downwards at the rate of about 19 degrees centigrade per kilometre. If this rate persisted all the way to the centre, the temperature there would be over 100,000 degrees. Almost certainly it cannot be so high and the rate of increase must die away at depth. Guesses of the actual temperature at the centre of the earth are between 2,000 and 6,500 degrees.

Locally and nearer the surface there are other sources of heat, especially radioactive heat. Most of the radioactive heat-producing minerals are thought to occur in the crust. Heat is also generated when rock surfaces move against one another, and in earthquake zones it is this heat which is generated when a quake takes place. Moreover, when a quake occurs the pressure within the rocks is released and melting begins to generate the magma for an active volcano. The molten material moves up the earthquake-jarred fault or crack to erupt but it is essentially a local 'hot spot'.

We can summarise the properties of the layers within the earth as follows. The outermost shell, the crust, is very heterogenous, being composed of granitic rocks in the continents and basaltic rocks under the oceans. Beneath the Moho lies the upper part of the mantle of iron and magnesium silicate (peridotite) to a depth of about 1,000 kilometres. The lower mantle extends to about 2,900 kilometres and is also made of iron-magnesium silicate material which because of the great pressure is changed to a very dense form. So tremendous is the pressure that the atoms within the iron-magnesium silicate are perhaps pushed together in the rather simple way that is found in minerals with cubic crystals. These materials tend to yield under stress, like malleable gold. Then comes the outer core, probably a nickel-iron alloy to 5,000 kilometres down. From here to the centre is the inner core, solid nickel-iron and very dense indeed.

Very recently a new subdivision of the earth has been proposed by scientists concerned with the movements in the outer parts. Refined seismology has located a zone between 50 and 250 kilometres down where earthquake waves travel only at much reduced speeds; it may be a region where major movements take place. The name asthenosphere, meaning literally weak sphere, is given to it. Above is the lithosphere—the sphere of rocks, while all that is below the asthenosphere is known as the mesosphere, the central sphere. The need to give names to these layers is apparent when it comes to discussing the origins of rather large lateral movements in the lithosphere. The asthenosphere seems to offer a means whereby deep-seated movements in the earth can be translated to movements of the lithosphere. It could be the cog between an inner machine and the outer wheels driving the geological cycle on which life itself is perched.

The great interest in earthquakes necessarily shown by the Japanese and Americans has recently become focussed in part upon the distinct possibility that earthquakes associated with the movement of the crust along particular fault lines occur at rather regular intervals. San Francisco as we know, for example, is situated on a major fault line, the San Andreas Fault. It seems that a major tremor occurs about every sixty years or a little more. In 1909 San Francisco was devastated by a major shock; another is due at the moment. In sixty years or so enough strain is built up along the fault to cause one block to move relative to the other. The Pacific side moves north-westwards a few centimetres with each quake. There is clearly a continuous force at work deep below which is bent on displacing one segment of crust relative to its neighbour. If only the fault plane could be lubricated to allow the one face to glide across the other the violent quakes might cease. There is a proposal to pump water into the fault zone in the hope that it would in effect act as a lubricant and prevent the faces locking for periods while stress built up.

Beneath Japan and the Indonesian islands very deep-seated earthquakes occur frequently and perhaps with some regularity in zones dipping under the islands from the Pacific side. The zones are called Benioff zones after a well-known American seismologist and they reach as far as 500 kilometres below the surface. In these zones there seems to be a constant pressure of some kind beneath the surface. The lithosphere jolts and creaks at fractures and along the deep earthquake zones. If the movements are indeed very frequent they may accomplish a great deal in the course of geological time even though the individual spasms of movement are very small.

Then there is the seismic activity of the mid-oceanic ridges. It seems perhaps not too odd that there may be earthquakes at or near the margins of continents, but in the centres of the oceans, where all is geologically rather uniform it was somewhat unexpected. But the ridges themselves were unexpected. Their discovery came as a surprise. We can be sure that with so much seismic and volcanic activity going on along these strange suboceanic zones there is something rather significant and fundamental in progress. There is more to be said about this in Chapter IX.

A Magnetic Personality

The popular belief that the core of the earth consists largely of nickel-iron is probably true. Not only would the evidence of meteorites suggest this, and the calculated density of the inner depths, but the presence of the earth's magnetic field is further support.

Mariners and navigators in Europe have known about the magnetic compass for nearly a thousand years, and until the time of Elizabeth I many of them believed that the North Star or the Great Bear constellation or some other feature in the skies was the source of the attraction for the needle. It was known that when hung vertically the needle dipped towards the north and that the angle increased as the needle was carried to more northerly latitudes. William Gilbert, physician to Queen Elizabeth I, showed that a sphere of magnetite (lodestone), being naturally magnetic, enabled him to duplicate this phenomenon. He concluded, correctly, that the earth itself acts as a great magnet.

This idea is, however, only a useful analogy. In several ways the comparison with a simple north–south bar magnet (a dipole) is only approximate. The lines of magnetic force are distributed in an irregular way, and fluctuate irregularly from time to time. It is well known that the magnetic pole does not correspond to the geographical pole but moves around continually over a wide area. There are in fact very few places on the earth where the magnetic needle points to true geographic north. This would make life very difficult for navigators were it not for the fact that measurements of the intensity and direction of the earth's magnetic field taken all over its surface allow the construction of charts with so-called isogonic lines, showing the differences from place to place.

Most of the earth's magnetic field is produced by something within the earth, but a small part is produced by effects above the earth's surface. They must be rather uneven or irregular happenings because the magnetic pole moves around. Since early in the last century its location has been kept under careful watch, but we have really no definite detailed knowledge about its wanderings previous to then. Not only has the position of the magnetic pole been changing by very small amounts from hour to hour but, it seems, the strength of the field has been changing too.

So little is known about the earth's magnetic field in detail that it is difficult to answer the most fundamental questions about its presence and behaviour in the geological past. For example, has the earth always had two, and only two, magnetic poles—north and south—and have they always been on opposite sides of the globe from one another or have they moved to and fro? Indeed, when did the earth first acquire a magnetic field?

Difficult questions, but happily they are not impossible to answer in part. Grains of the magnetic iron minerals, magnetite and haematite, present in basalts and sandstones, make these rocks in effect very feeble magnets. When the rock was formed each grain was magnetised in the same direction as the magnetic field of the time. In addition, the degree of magnetisation was proportional to the strength of the earth's field when the rock was formed. So each magnetite or haematite crystal or grain acts as a tiny magnet bearing the imprint of the earth's magnetic field at the time the crystals grew or the grains came to rest. If we can measure the strength and direction of this very faint magnetisation we have some information about the field when the rock was formed. And this is indeed possible because once the magnetism has been imparted to the rock it is very difficult to change it. Once magnetised, the rocks faithfully retain their magnetism for hundreds of millions of years.

During the last fifteen years or so there has been a palaeo-magnetic 'rush', a great effort on the part of geophysicists to get out and collect rock samples of as many different geological ages as possible. The top and bottom of each sample is marked, the direction of north today and the attitude of the rock in the ground. In the laboratory the samples are tested by highly sensitive instruments to measure their magnetic properties, the strength and direction of the magnetism that was incorporated in the rock when it was first formed, the 'fossil' or palaeo-magnetism. With enough of these samples it is possible to gain some idea of the position and strength of the magnetic poles of the past. Thousands of measurements are needed from rocks that are as nearly contemporaneous as we can tell and from widely separated localities, and from these measurements averages can be calculated.

From the study of palaeomagnetism some interesting and even astounding facts emerge. In the first place it has been discovered that in the large number of basalts formed during the last 7,000 years the north magnetic pole is indicated as having wandered around the Arctic Ocean basin pretty widely, but it does seem to have been centred upon the geographic pole, as though it were on some sort of leash. Apparently the magnetic pole never gets further than about 11 degrees away from the geographical pole, and during these 7,000 years there has been one, and only one, north magnetic pole.

When slightly more ancient rocks are examined the position of the magnetic pole varies widely, in fact so widely that it seems to afford some remarkable corroboration for the idea of continental drift. There is more to be said about this in Chapter VIII.

There was an unexpected discovery made while the palaeo-magnetism rush was on. It had previously been found that some rocks seemed to have north magnetic poles where the south should be and so were called reversely magnetised. Normally magnetised rocks are those magnetised as a rock formed today would be. Reversed magnetism was now proved to be not a random business produced somehow by a local event or change in a particular rock, but a world-wide affair affecting all rocks in process of formation at certain times. Geologists of the U.S. Geological Survey discovered that rocks going back in time as far as 4·5 million years are alternately normally and reversely magnetised. They distinguished several phases of re-versed polarity. Unfortunately in rocks older than 4·5 million years it is not possible to correlate accurately enough to be able to distinguish separate times when the magnetic field was suddenly reversed, but we know that it did occur. The periodically contrary character of the field seems to have been always part of the earth's magnetic behaviour.

The oldest rock found to have an original magnetisation is about 2,600 million years old. What happened before then is not certain, but in a theory about the constitution of the earth these facts all have to be accounted for.

William Gilbert, who experimented with a magnetite sphere in Elizabeth I's day, suggested that the earth is a permanent magnet. Popular though the idea has been, it seems to fail in that when iron or iron minerals are heated to over 500 degrees centigrade they lose their magnetism. Well, there is good reason to believe that at only a few kilometres below the surface of the earth the temperature is quite a lot higher than that. So the magnetism is not likely to originate down there. Does it then reside somehow in the thin outer skin or shell of the planet? There is, we discover, no concentration of magnetic material strong enough in the outer skin to produce a field as strong as that which clearly exists.

From this position geophysicists took a second look at the deeper parts of the interior. They had in mind the principle of the dynamo, the fact that any moving electrical charge sets up a magnetic field. If there was movement in part or all of the fluid core, which may well be of iron and have a metal-like atomic structure, a large magnetic field could be produced. All that is needed is some energy to keep the core moving. As long as it moves, the core excites a magnetic field big enough to encompass not only the solid earth but a considerable range of space beyond it. Of course, no one can say for sure if this is what has been happening in the fluid iron outer core of the earth throughout geological history, but given the intense heat of the smaller solid inner core or the energy from chemical differences between core and mantle, there is a possibility that this kind of motor–generator combination could exist. We imagine that in the earth the movement of the conductor is a sort of convection of the iron core, but not a rapid nor orderly one. Because of the effect of the earth's rotation, the convecting fluid follows swirling complex paths, rather than simple ones. These irregularities cause continual changes in the strength of the magnetic field and the positions of the actual magnetic poles.

In its simplicity this is an attractive theory, and both experimental geophysical and recent space research seem to confirm its validity. It accounts for both the overall orderliness and the irregularity seen in the earth's magnetic field. The rotation of the earth accounts for the constant close proximity of the magnetic and geographic axes, while the more irregular movement of the connecting currents in the core could be responsible for the incessant but limited wandering that does take place. The magnetic forces thus set up are felt not only on the earth but extend as a zone, the magnetosphere, 25,000 kilometres (40,000 miles) out into space. The magnetosphere appears to act as a kind of trap or filter for bursts of radiant energy and subatomic particles continually hurled earthwards by the sun, and it deflects away much of this solar wind.

A voice from the rear may at this point ask what significance the reversals of the magnetic field may have in all this, and if the reversals are instantaneous. For the answer, we should turn for the moment to the sediment accumulating on ocean floors, sediment in which the magnetic reversals are recorded almost as infallibly as they are in contemporary lavas. Cores taken from these sediments by deep water oceanographic research teams show that at the reversals some of the smaller forms of life become extinct. Presented with this evidence, the oceanographic biologists thought that perhaps at the moment or period of reversal there is for a while virtually no magnetic field to hold back the cosmic rays, the strongest radiation from the sun. Possibly, then, the sudden bombardment by cosmic rays causes extinction of this species or that. Even allowing for the brief reduction of the magnetic field, however, the atmosphere alone probably affords enough protection from such a bombardment, and certainly the ocean does, so that marine life would be little discomforted below the surface. But the sediments also offer another clue, they contain an extra abundance of tiny meteorite particles, known as tektites. Tektites are minute fragments of melted rock produced when a meteorite smashes into the earth. It has been proposed that when a very big meteorite strikes the earth the jar makes the earth's core dynamo go into reverse and the direction of the magnetic field is changed round. Plenty of very large meteorites are known to have hit the earth and some have left gigantic craters. It seems quite possible that the large ones could have upset the effects of the terrestrial dynamo outlined on these pages.

Other planets which have magnetic fields, Jupiter is one for example, are thought similarly to have liquid cores. Where there is no magnetic field the planet may be solid throughout or lack the essential conducting element, iron. If these planets have liquid cores, iron cores, they also need a source of energy—an engine to keep the dynamo running. The movement necessary could be provided, we think, in three possible ways. If, in the case of the earth, the inner core were still to be growing the heat given out during the change from liquid to solid could provide the energy to move the outer core. There are, however, some fairly substantial objections to the idea of a growing inner core. Tidal effects of the sun and moon could cause effects within the earth that kept its internal engine turning. Thirdly, while it's generally agreed that the elements with large atoms, such as the radioactive uranium and potassium, are most abundant in the crust because they are squeezed out of the deeper realms, there is the possibility that radioactive heat may be generated in the inner core where a high proportion of the very heavy element, uranium, could be present. As far as other members of the solar system are concerned, the origins of their magnetic fields are even less certain than those of the earth, but at least there now seems to be a not too creaky hypothetical model for this particular planet.

Shell Within Shell

At some stage in the evolution of our planet its internal anatomy of core, mantle and crust must have begun to take shape. Opinion seems to be general that it must have begun early on and that from what may have been a rather uniform globule or mass of hot material the heavy metallic elements soon began to gravitate in towards the centre. Perhaps over the first 500 million years of its history the earth experienced this inward settling of the nickel–iron and some silica droplets and particles separating out from the silicates and other lighter materials that were left as an outer layer. It is difficult to suggest what the surface of the earth was like at this stage. Scorching hot, with volcanic eruptions and great tracts of molten rock giving off gases and fumes at a great rate, it would have appeared inhospitable beyond belief.

This heating and boiling of the crust may have reached its peak only after the core had taken shape. The lighter materials bearing with them most of the radioactive elements in the earth would have been left behind near the outer part of the planet and between 3,200 and 2,800 million years ago their heat production would have been three or four times what it is now and it would have been generated by the elements thorium, potassium and rubidium as well as by uranium.

The material in the outer core may have begun to move, even to convect in some manner, as soon as it had accumulated in sufficient thickness. Even after it had assumed much of its bulk there may have been a longer period of very slow growth of the core as further particles of nickel–iron made their way inwards from the mantle.

In its youth the mantle itself was probably different in character from its present state, with only a very thin crust of lithosphere above. Indeed, being very hot and fluid the mantle may have had a much thicker asthenosphere than is present today. What movements there may have been in the asthenosphere we can only guess. Some geophysicists have favoured rather large numbers of convecting cells within the mantle, which could have led to complex patterns of movement being transmitted to the lithosphere. As time has passed the number of convecting systems within the mantle has dropped and the effects at the crustal exterior have become simpler and more linear.

Be that as it may, there is now new evidence from West Greenland of rocks as old as 3,980±170 million years. They are very metamorphosed rocks and granites, showing sworled and whispy structures more involved and complex than any that have been produced on a wide scale since. These structures are unlike others anywhere else. All these most ancient rocks are known over an area wide enough to suggest that the weird patterns into which they are thrown seem to be consistent with the idea that the crust, of which they formed a part, was contracting. The shrinkage was general, occurring in all directions. This, it is thought, might result from physicochemical changes taking place within the mantle itself. Breakdown of the ferromagnesian minerals, with perhaps the release of water, might involve a small but important loss of volume in the mantle and in turn the crust would respond.

Rocks older than 3,600 million years are extremely rare indeed, but between 3,600 million and 2,600 million years ago the crust of the earth seems to have acquired both granitic and basaltic rocks. We cannot say whether the crust prior to 3,600 million years was predominantly granitic or basaltic because there is only such a miniscule part of it left for us to examine. Within the next 1,000 million years, however, granitic crust in large volume seems to have been present, and adjacent to it or upon it eruptions of basalts took place. The evidence for these eruptions we shall meet in the 'greenstone belts' described in the next chapter. The earlier greenstone belts were rather small; the later ones were much larger and by that time the crust would seem to have been stable enough to develop depressions in which many thousands of metres of erupted volcanic materials could accumulate.

This is the picture we reconstruct from the evidence within the continents. The oceans yield no rocks anywhere approaching these great ages. It has been suggested by one geologist that when the greenstone volcanic belts were taking shape the originally thinly spread lightest and outermost skin to the earth had begun to gather into several local and thickened masses, proto-continents. Continued eruption of material from the mantle through the lithosphere brought small but significant additions of light granitic material to the outer surface. Now and again vast areas of the crust would founder and melt as radiogenic heat accumulated or basalt floods smothered the thin crust. The melting served further to separate the light granitic magma from the heavier basaltic material.

The lithosphere eventually became thick enough and cool enough to resist such foundering and the granitic matter lay in large 'bergs' on top of the basalt. Continents of a kind were formed.

Volcanoes, earthquakes, magnetism, meteorites, planets and the sun all provide a bewildering array of facts that must be incorporated into our ideas of the nature and history of the earth. The earth is clearly far from being a simple round lump of rock within an envelope of air and water. Most of its surface features give little indication of what lies at great depth, though volcanoes and earthquakes help. From the study of these and indeed of all the processes that seem to operate from below the earth's surface emerges a reasonable idea of what exists below. It is improved by ideas derived from extra-terrestrial objects. There is, however, no concrete evidence to suggest that the interior of the earth has really changed substantially since the crust was first formed. With so much obvious heat and energy not far below the surface even now, the earth must have had a stage very early in its history when the solid crust was yet unformed. We can do little more than theorise about the events of that distant time; almost no trace of them remains.

Although earth seems to have perhaps more features in common with other planets than was thought fifty years ago it remains a unique planet. What continues to be emphasised by recent researches is the dynamic character of our globe and the tremendous variety of forces at work around and within it.

CHAPTER IV

Back to the Beginning

OF COURSE it had to begin somewhere and at some point in time; man always seems to have been sure of that, although his explanations of how and when have changed. Early ideas on the origin of the earth seem delightfully uncomplicated and in most cases involve a divine act or two. As more thought has been given to the matter, men have felt that they may be getting nearer to the truth, whatever that may be. Scientists have had to explain more and more facts, embrace bigger and more complicated hypotheses involving ever increasing quantities of matter, time and space. For most people the whole business has got out of hand and most geologists admit that they prefer to keep their eyes respectfully on the ground rather than contemplate the insignificance of the earth in the cosmos. But all is not unintelligible and the welcome news is that enormous though our galaxy may be and incomprehensibly large though the cosmos may be, the increase is indeed finite. There is a limit to what is there even though, say some, there may be no measurable limit to the space it occupies. As things stand today, the cosmologists claim to observe galaxies drawing away from us at nearly half the speed of light. The speed of light (297,000 kilometres per second) is perhaps the limit which delineates the extent of the observable universe. A ray of light released anywhere in the cosmos may travel infinitely far, going round and round in curved space. The radius of this curvature is calculated to be about 13 billion light years.

The investigations of the last few years have brought to our attention the fact that the universe is not a very tidy place. Apart from the conspicuous lumps of matter such as stars, suns and galaxies, there are rather less obvious nebulae of dust and gas and even from 'blank' areas in the sky comes evidence of scattered thin and local clouds of frozen gas and fine dust. Nearer at hand, within the solar system, there is a constant emanation of particles streaming out from the sun, known as the solar wind. Its effects are still unfathomed but they might, over long periods of time, be significant. Much of space elsewhere may experience solar winds from other suns.

This is heady stuff and seems to offer little of comfort to the earth historian, suggesting perhaps that at other points in time and space events and conditions similar to those experienced on this small planet, revolving about a small star in one of infinitely many galaxies, may exist. Earth, nevertheless, is to all intents and purposes a unique item in the universe. After all, we live on it, are affected for better or (as our ancestors thought) for worse by its simple satellite and are dependent upon that fountain of energy at the centre of the solar system, the sun. The rest is remote, not unimportant but remote.

Perhaps it was this obvious remoteness, coupled with the periodic movements of the stars, planets, sun and moon that appealed to early man. He thought confidently that it all revolved about the earth. The night skies were mapped out into fanciful patterns, the constellations. The planets were ultimately all recognised and the comets and meteorites recorded. Supernovae, sudden blazes of starlight lasting for a few weeks or months, were watched and discussed.

Each civilisation has had its astronomers—Chinese, Greeks, Syrians, Egyptians, Mayans, and even the ancient Britons with their huge but astronomically accurate building, Stonehenge. Modern astronomy traces its descent through the Greeks, Galileo and Newton. Sir Isaac Newton, known universally for his law of gravitation, took more than twenty years to mull over and eventually publish his book on its ramifications. Today's schoolboy meets these soon enough but when Newton first outlined how gravitational and centrifugal forces act within the solar system he was at the very frontier of knowledge at that time. Much that had been puzzling in astronomy began to make sense, but the earth was no longer the centre of the universe.

The laws of physics and chemistry have since been called upon to help unravel some of the mysteries of astronomy—and investigations into such fields as optics, electromagnetism and atomic physics have been largely prompted by the needs of astronomy. The earth has been relegated to the role of a small satellite to a small to medium-sized star, one of many in a galaxy which stretches across our sky as the Milky Way. At unimaginable distances are other galaxies, spiralling discs and bands of innumerable suns and solar systems.

The further away they are the less, it might be said, is known about them. To look beyond the solar system for clues about the origin and evolution of our own planet is now not unrewarding for the geologist, but events and conditions beyond this corner of the cosmos seem somehow unreal. The birth and history of the solar system involve concepts large enough for the average earth scientist. His knowledge of sun and planets is very fragmentary but most scientists have a firm belief that the fundamental laws of nature have not changed with time. There is, however, really no proof that, for example, the speed of light or the constant of gravitation have not changed in time. It would make the whole process of unravelling geological history, not to mention cosmic history, much more difficult if such things were not constant.

From spectroscopic analysis of the light of the sun and stars we have a pretty good idea of the kinds of elements and the proportions of them present in these heavenly bodies. But, as we remarked above, there are also clouds of other materials in space. Both luminous and black opaque clouds of gas and dust are observed at great distances out in space. Their ages remain as mysterious as their compositions.

As far as time is concerned, the age of the earth is fairly well agreed as 4,600 million years, and the moon seems to be much the same. Meteorites give radiometric dates which suggest that the planetary body they came from is—or would be—about that age too. What existed or happened before then is unknown. It is sufficient here to regard it merely as the beginning of creation; the dawn of the first day of Genesis.

A Celestial Family

The solar system seems, on the other hand, to be a tidy little celestial machine, behaving with commendable, if not perfect, regularity. At the centre is the sun, containing more than 99 per cent of the mass of the system and measuring about 1,382,000 kilometres (864,000 miles) in diameter. The planets are strung out around it in elliptical orbits revolving on their axes and rotating but only slowly, and not in a uniform manner. Being a ball of hot gas the sun appears to swirl round faster at the equator than near the poles. At the equator it makes a revolution in just over 25 days, but at the poles it needs 33 days. Sixty per cent or more of it is hydrogen, while helium is perhaps half as plentiful by weight.

In such a large mass of incandescent gas this swirling movement is very rapid. All the calculations indicate that the centre of this system should be spinning with a rotation every ten hours or less which is a great deal less than the observed 25 days or so. So the hypotheses which hold that the sun and the solar system originate from the condensation of a swirling mass of gases and dust, a nebular, are regarded as interesting but in the end do not adequately explain the movement of the sun and its attendant planets.

Casting about the skies, it is found that other old stars like the sun are also rotating slowly; many new stars on the other hand spin round very rapidly. There is obviously some force at work which is unlike those known on earth. Could there once have been some form of braking or slowing interference by other-planets or stars?

The inner group of planets consists of tiny Mercury (4,800 kilometres in diameter) close to the sun, then Venus, Earth and Mars in outward succession. Jupiter and Saturn are the two largest of a quartet of giants which also includes Uranus and Neptune in order from nearest to farthest from the sun. At the edge of the system is Pluto which is thought to be rather like the inner terrestrial group.

The smallest and innermost planet, Mercury, is sometimes seen as a bright spot in the sky but it is so close to the sun (57 million kilometres) that it is difficult to study. In the evening Mercury sinks from view almost as soon as the sun does and observations at low angles through the earth's atmosphere are notoriously unreliable. Nevertheless there are several interesting things known about this elusive planet. The Mercurian year lasts a mere 88 earth days, but a Mercurian day is about 59 earth days. At Mercury's midday the temperature rises to about 415 degrees centigrade while at midnight it has cooled off to a comfortable 21 degrees centigrade. The atmosphere is a thin affair of gases rising out of the interior and may be too heavy to swirl quickly away into space. From the little that is known, the surface of Mercury may be desert, perhaps, rather like the moon's. The density of this little planet is 5·1, rather less than the earth's 5·52 and perhaps similarly due to a high proportion of iron in its core.

Venus is only very weakly magnetic, which seems to indicate that either it does not have a liquid iron core like the earth's or that the core is very small. Some astronomers suggest that its iron is perhaps scattered throughout the body of the planet, even staying close to the surface and oxidising there until no oxygen remains in the atmosphere.

Both Venus and Mars fascinate because scientists have thought some forms of life could possibly exist upon them. Venus is the brightest star in the night sky and comes nearer to earth than any other planet. Its orbit lies between that of Mercury and the earth at 67 million miles from the sun and its surface is obscured by a permanent

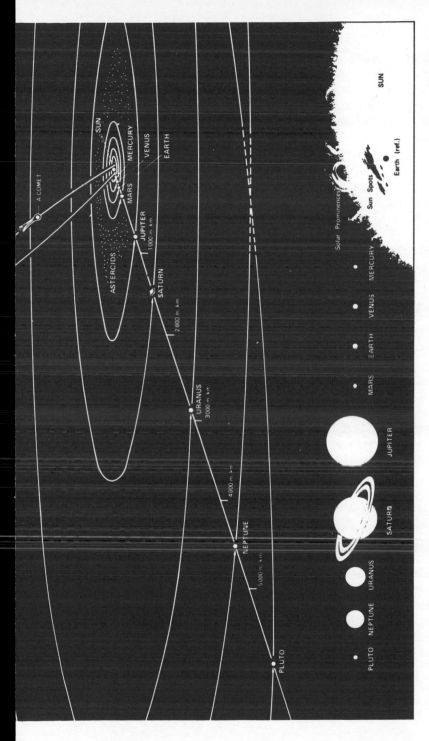

Figure 4–1   A diagrammatic summary of the solar system. All the planetary orbits except those of Pluto and Mercury lie in nearly the same plane as that of the earth. Recently it has been suggested that there may be a tiny tenth planet far out beyond Pluto, but it remains so far undetected. The sun and the known planets are drawn to scale in the lower part of the figure.

(Reproduced by courtesy of Artemis Press.)

layer of clouds. This cloud screen prevented the actual size of the solid planet from being measured directly but the properties of the Venusian atmosphere suggest that the planet is only a mite smaller than the earth, with a diameter of 12,100 kilometres (7,560 miles). Its density (5·2) is only a little less than that of earth. Venus has been scanned by the American rocket Mariner II which transmitted signals to the effect that the greater part of the surface of Venus is almost red hot, so the idea of Venusian life seems to have been dealt a serious blow. This high temperature is probably maintained by 'the greenhouse effect'. In our greenhouses sunlight warms the soil and air and they in turn radiate heat back. However, because their temperature is relatively low they only emit long waves of radiation which, as it happens, do not easily pass through glass. Their heat energy is trapped and pushes up the temperature. In place of glass, of course, Venus has carbon dioxide and extremely tiny ice-crystals on high with water vapour near the planet's surface.

Radio astronomy has penetrated the dense Venusian atmosphere to reveal that the planet rotates in the opposite direction from all the other planets, except Uranus. The Venusian day is about 250 earth days and its year is 224·7 days, which means that sunrise occurs only once every 118 earth days. Midday temperature could reach 720 degrees centigrade and night-time temperatures go as low as 320 degrees centigrade, perhaps less in polar regions. There is such a high atmospheric pressure on Venus that water could possibly exist locally in relatively cool spots, but it would be super-heated stuff in comparison to earthly water.

Further discoveries by the radio and radar astronomers indicate that Venus has surface mountain systems and larger smooth areas of dust or pulverised rock. Possibly there are wind-blasted surfaces to the rocks, like those in the most arid deserts on earth. All in all, the prospect for life on Venus is not good until it cools down a little.

Mars is rather a different matter and almost everyone has his own ideas about Martians and life there. This is a planet which, only a little further from the sun than the earth, 227 million kilometres, has an atmosphere, with clouds, white polar caps, varied topography, and seasonal change in appearance very like the earth's. Mars takes 687 earth days to orbit the sun and its axis of rotation is not perpendicular to its orbit but is tilted at about the same angle as the earth's.

Two small satellites circle very close to Mars. Deimos, only about 5 kilometres in diameter, moves at some 24,000 kilometres up, while the 8 kilometres-diameter Phobos cruises at a mere 9,275 kilometres.

The Mariner IV spacecraft was able to view numerous circular shapes that may be impact craters on the Martian surface. Close to Mars lies the asteroid zone, where what was thought to be the debris of a collapsed planet or planets, revolve around the sun like so many chunks of shrapnel. Many of them are huge but there must be countless others, meteorites, too small to be seen from earth. Most of the asteroids remain in proper orbit, but some are known to wander eccentrically. The earth, the moon and Mars have all suffered impacts from meteorites, with local devastating effect. Poor Mars, however, has suffered the worst bombardment but Martian erosion seems to remove the scars. The thought that in the past Mars had a watery atmosphere rather like our own has crossed the minds of several modern researchers but it is now discarded on the evidence of the Mariner spacecraft.

The pressure of the Martian atmosphere is only about 1 per cent of earth's and carbon dioxide is the principal gas, nitrogen also being present and a tiny percentage of oxygen and water vapour. There is virtually no 'greenhouse effect' in this atmosphere and the place is not only dry but variable in temperature from a noonday 21 degrees centigrade to 25 degrees centigrade to a night-time −85 degrees centigrade at the equator.

The lack of oxygen and water vapour in the Martian atmosphere may be linked with the fact that Mars has no detectable magnetic field. A density of 4·0 and the absence of magnetism may indicate the lack of an iron core, the planet's iron being scattered throughout the whole body. At or near the surface this iron may have reacted with water vapour and oxygen. Characteristic of such a rusty surface would be the orange-red colour that is so distinctive to Mars.

At an earlier time Mars may have had more water, together with ammonia and methane in its atmosphere and the climate may have been warmer. The conditions then would have been possibly suitable for life of a sort to evolve. Once it had appeared it could have been as resourceful as life on earth in adapting to the slowly changing conditions on the planet. There could, in theory, be other patterns of chemistry upon which life of a kind could be based, and first development of life there could have happened before life was manifest on earth. Its evolution could have been faster and very dissimilar from that here.

Jupiter is the biggest and perhaps strangest planet, the nearest of the giants, at 772 million kilometres from the sun. It has a diameter more than 11 times that of the earth, has 12 satellites and contains more matter than all the other planets put together. Big it may be, but Jupiter possesses a very low density, only 1·33 compared to that of 5·52 for the earth. The reasons for this are that while Jupiter has an atmosphere, thousands of kilometres thick, of ammonia, methane and hydrogen, the planet itself is largely composed of hydrogen and helium with only minor amounts of other elements. All these are under the tremendous gravitational pull of the planet. Jupiter has a small magnetic field which may be produced by movement in a small iron core or by iron being moved about by the gases on the planet.

The abundance of these light elements on Jupiter (and their scarcity on earth) show that the planet has changed little in composition since it condensed from gas and dust 5 billion years ago. Only when, as some astronomers believe, the sun grows much brighter will the temperature of Jupiter's atmosphere rise enough to let the hydrogen and then the helium escape.

The surface of Jupiter's atmosphere seems to have a latitudinal banding as it blows incessantly in a westerly direction. There is no nitrogen in this atmosphere but ammonia and methane have been detected and water should also be present. In the southern hemisphere is the gigantic Great Red Spot moving rather slower than the band around it. It is about 12,500 kilometres by 40,000 kilometres and seems to change colour and shape from time to time. The most likely explanation of this unique discolouration is that it is the top of a relatively stagnant column of gas which is perched on some extra large topographical feature in the planet's surface.

Jupiter has another extraordinary feature useful in our debate about the history of the solar system. It seems to be actually giving out heat, despite its very low temperature. Thus it resembles a star rather than the other planets. The centre of this planet is not hot enough to permit the thermo-nuclear reaction that heats the sun so it must have only gravitational heat to give off. Which means that perhaps Jupiter is still contracting as the sun did long ago.

More than one astronomer and several biologists believe that complex organic molecules may be built up within Jupiter's atmosphere and most, if not all, of the elements for life exist within it. A form of life developing within the medium of a dense atmosphere rather than in water is not to be rejected out of hand.

Saturn comes next, at 1,463 million kilometres from the sun, and is, with its distinctive halo, an even colder and less hospitable planet. It is only a fraction smaller than Jupiter and has a similar deep atmosphere of a like composition. There may be atmospheric disturbances as on Jupiter. What the surface is like beneath this pall of cloud is not known. Saturn (120,000 kilometres in diameter) has long attracted attention because of its distinctive halo, which is found to consist of tiny particles, probably ice crystals. Although the halo rings have a diameter of 273,100 kilometres they are only a few centimetres thick. There are also nine satellites.

Uranus, the third outer giant and 48,000 kilometres in diameter, circles at about 2,864 million kilometres from the sun. It is very similar to the other two with a low density and even lower temperatures. Neptune is very similar again but is, at 4,480 million kilometres from the sun, over twice as far from the centre of our system.

At the very edge of the system is a planet that was unknown until 1930; Pluto is an estimated 5,888 million kilometres from the sun. Even now relatively little information has been obtained about Pluto. It may be cold or it may have some internal heat of its own. Even the presence of an atmosphere is uncertain and the composition of the planet is unknown.

Thus the solar system contains planets that are distinctive in their own ways, and in the midst of them is our own, the planet Earth. Clearly we are learning about our nearest celestial neighbours at a faster rate than ever before, and although scientists may argue about many of the facts mentioned above, they do recognise that the solar system is indeed a system and not a haphazard collection of satellites spinning in random orbits about a star.

A Sun is Born

Early astronomers had of necessity a simple idea of our solar system and the theory needed to explain it. Their ideas were at the forefront of science of their day. With so much more to paint onto our scientific canvas it needs teams, rather than individuals, to work out the master pattern and the sequence of events that produced it. Astronomy has become a highly technological branch of science, its data complex and its concepts difficult. It offers us no simple unequivocal theory about the origin of the earth.

The birth of the earth was not an isolated event. Its process may have been unique but it must have had much in common with that of the other planets. Since the formation of the solar system, all the planets have evolved. They have changed with time, perhaps in composition, size and other physical attributes. Cosmic events as well as those triggered off within the planets themselves have affected their evolution. At some stage or another, life of a kind may have made its appearance on planets other than earth.

As early as 1745 Georges Louis Leclerc, Comte de Buffon, the famous French naturalist, suggested that another body had hit the sun, knocking from it the spinning gas and matter that became the planets. Buffon suggested a giant comet; other writers have put up ideas that involved stars passing by or colliding and even a twin to the sun, all of which have since swung off into space. These ideas also are intriguing, but they fail to explain the dynamics that are required. One such theory produced by Sir James Jeans in 1901 attempted to improve on Laplace's idea by visualising a great wing of solar matter being drawn out of the revolving sun by the attraction of a large star wandering by. Fragments of this wing condensed into the present planets and asteroids. Unfortunately for Jeans' hypothesis it is now known that a wing or column of solar matter, drawn out from a parent body with a temperature of a million degrees or more, would simply explode in such a violent way that no planets could condense in a neat disc about the parent star. Other physicists, the German Carl von Weizsäcker for example, have imagined the angular momentum of the sun transferred to the planets as turbulence developed in the Laplacian nebular disc. This idea also has had to be abandoned. It seemed for some years as though no progress could be made in solving this fundamental problem. Then astronomers in Scandinavia and North America have recently toyed with the idea that a strong magnetic field might have been a major factor in the way the planets formed. These notions prompted the Cambridge astronomer Fred Hoyle to suggest that as the early sun collapsed inwards under its own developing gravity, magnetic forces acting on the hot gas drew it out in threads of columns, spinning in a disc.

The ability of the sun to spin rapidly upon its axis was modified by the enormous magnetic field of whirling gas and carried out across the disc. According to Hoyle, only 1 per cent of the solar mass would be needed to constitute the disc and as the material cooled, the outgoing gas would carry with it small solid particles. In time the solids and some of the gases would collect into the various planets. This theory involves some extremely difficult calculations and calls upon processes involving very hot gases in which electric currents and magnetic fields have play. These processes are quite unfamiliar affairs and were not visualised until a few years ago when astronomical data and theoretical physics revealed them as a promising agency in planet formation. Something of this concept of magneto-hydrodynamics, as these magnetic processes are called, was initiated when the Van Allen radiation belts were detected extending about the earth not many years ago. It is known that charged particles can move easily *along* the lines of a magnetic field, but not *across* them. The movement of particles along the magnetic field is controlled not only by complicated hydrodynamics but by electromagnetic forces as well. In combination these forces may have determined how matter spun from the sun, moved and coagulated.

The idea of a primeval sun and its magnetic field of swirling gas and solid particles is reasonably acceptable, but what of the processes in the disc that were to produce such a peculiar family of planets? Truly it is a weird assortment of heavenly bodies circulating round a sun consisting primarily of hydrogen and helium and with carbon, nitrogen and oxygen making up only 1 or 2 per cent of its mass, and all the other elements present as mere traces. Many of the known chemical elements have been detected on other planets but the proportions differ in each of them. The terrestrial planets Earth, Mars, Mercury, and Venus increase in density from the sun outward. Earth's density (5·52) and seismic data, as we saw in Chapter III, suggest an iron core. Nickel-iron occurs in many meteorites and may thus have been present in a now disintegrated planet or may represent cosmic matter that never completely jelled into a planet. In the other planets we might expect more or less iron to correspond to the higher or lower densities that they possess. What, in fact, has caused the elements to become distributed among the planets in the peculiar way they have and why have the outer giant planets, Neptune, Saturn and Uranus, such small densities?

Among the puzzles presented by the chemistry of the earth is the abundance of light elements. In any condensation from a hot gas cloud the hydrogen and other light elements would surely have been driven off. To account for the presence of these light elements the idea of accretion from a cold mass of dust and frozen gas particles has been proposed. We know that such clouds certainly exist in space. They can be seen as opaque blots in front of distant nebulae. Some of them have at least as much matter as is in the sun spread out over a space equal to that from the sun to the nearest star. Dr Lyman Spitzer of Princeton University has produced the novel but rather convincing idea that such clouds can be affected by light from neighbouring stars. In effect the dust particles could be swept together by persistent starlight. Eventually the clouds would be sufficiently compressed by this means to collapse under their own gravity. The compression allows the cloud to heat up and, eventually when colossal temperatures are reached, thermonuclear reactions begin. The 'dust-cloud' hypothesis has received much support recently and space researchers favour this 'cold accretion' idea for an origin of the earth.

It has been pointed out that the elements can be divided into three natural groups in terms of abundance and characteristics. There is a *gaseous* group, helium and hydrogen; an *icy* group which make up compounds such as ammonia, methane and water and freeze at moderately low temperatures; and an *earthy* group comprising the rest of the elements except the heavy inert gases. The earthy group makes up the bulk of the terrestrial planets but only small parts of the giant planets, where the icy group prevails. All along, it has been thought reasonable to believe that the original source of building material for the sun and all the planets has been the same. Perhaps the inner four planets, the terrestrial group, has lost a vast quantity of the lighter icy group of elements. If this were so, there must have been a mass of solar material about 500 times greater than the existing terrestrial planets. The condensation process would at best need about 3 per cent of the present sun's mass to provide the original material from which to derive the planets and comets—probably it was much more. How the gases were removed and how the condensation into planets took place is not known. The difficulties of removing so much gas from these protoplanets are so enormous that at least one scientist has sought to tackle the problem of the origin of the planets from a completely different angle.

Harold C. Urey, an American geochemist, took up the point that earth and the meteorites have lost significant quantities of volatile material—the icy elements were lost except where they combined with earthy elements—and he suggested that the matter from which the solar system evolved was first not hot, but very cold. Only later did aggregations and clots of this cosmic dust and gas develop high temperatures, possibly by radioactivity. As the material hurtled along in space the aggregated lumps collided, fragmented and reaccumulated. In doing so the various elements segregated into combinations such as silicates, gaseous elements, etc.

The earth's atmosphere is unusual compared to those of the other planets because the heavy inert gases are very rare. It seems most likely that any early atmosphere that was attached to the new-born earth was soon stripped or blown away by heat and the solar wind. The present atmosphere and hydrosphere are perhaps then of secondary origin, derived from within the earth. Mars has very little atmosphere and Mercury almost none, but because these little planets have such low forces of gravity their atmospheres are easily lost. The situation on Venus is still not clear. Before a theory for the origin of the earth and planets can be accepted it must account for the differences between the compositions of these atmospheres.

In 1964 F. J. Whipple presented to the U.S. National Academy of Sciences an account of the possible formation of the solar system, starting with a protosun that was a cold and relatively dense cloud of hydrogen and dust. There was turbulence in this cloud and a rotating motion at about the same rate as that for the galaxy, i.e. about one revolution in 100 million years. Because of supernova explosions nearby and eddies in the Milky Way, or because of the influence of a nearby larger cloud, the cloud became gravitationally unstable. A central clot of matter became segregated, attracting particles to it from the cloud and developing a magnetic field. The process grew faster until the cloud was virtually collapsing and the central core growing rapidly while speeding its rotation and sending out stronger lines of magnetic force. The temperature rose, being highest at the centre. By then the thermonuclear reactions which give the sun its immense energies began to make themselves felt. We have at this stage the nebular disc around a glowing central sphere that Laplace imagined 200 years ago. As the disc, consisting largely of hydrogen, spun out its material into planetary orbits, the temperatures fell rapidly because of radiation. At this stage the sun might have been so big as to engulf the present orbit of Mercury. The outer planets, Jupiter and Saturn, were soon formed by rapidly cooled nebular material while Uranus and Neptune were produced by the accretion of lumps of solid water and dust, only a little of which found its way into Jupiter and Saturn. Uranus must have collected a rather large chunk of cosmic matter or a small protoplanet which caused it to tilt its orbit out of the plane of rotation occupied by the orbits of the other planets. Pluto's highly inclined orbit (17 degrees) may be due to similar collisions, or due to quite different reasons. Between the growing outermost planets there were vast cosmic snowstorms.

On the outskirts of the early solar system the cold matter of the nebular disc may have aggregated as many small masses of dirty, icy material, giving a belt of comets. From time to time passing stars attract some of the comets out of their domain beyond Neptune, even to this day. Many of these comets must have been captured by the planets as time went on, some reaching towards the sun, others thrown out into interstellar space. Collisions eliminated many and others were thrown into very elongated orbits, some returning near to the sun at long intervals.

As the inward collapse of the protosun continued and the larger outer planets took form, the magnetic lines of force within and around the sun grew stronger and collisions of aggregating lumps and dust clots within the cloud became more violent. For a long time the sun had been contracting but when this stopped the release of solar energy rose sharply. The embryo terrestrial planets were by then formed from growing aggregates of dust warming up in the developing solar heat. With the increasing heat, the emission of gas or solar wind from the sun also increased, eventually becoming strong enough to blast away the gaseous materials of the terrestrial planets. In the meantime these particular planets became completely molten, largely from the heat generated within them by gravity and radioactivity.

The Years Begin

The earth now evolved from being a uniform fluid mass to being a rather more complicated body. It was beginning to segregate into concentric shells of matter, the innermost being the densest and produced by the inward migration of the abundant particles and droplets of nickel–iron. The outer parts were thus left relatively iron-poor and silicate-rich, forming the mantle and crust. Enveloping the thin new crust and growing with volcanic additions was a hot atmosphere of water vapour, nitrogen, methane, hydrogen and small amounts of other gases. Earth had by this time become different from other members of the solar system. Although it may have been scarcely recognisable as earth, it was acquiring the major concentric structure that has been its characteristic ever since.

No doubt volcanic activity and crustal disturbances on a gigantic scale took place with great frequency and ferocity. To the original atmosphere of cosmic gases new gases and vapours were added from volcanic punctures in the crust.

In time the temperature dropped sufficiently to allow boiling hot rain to fall. Rock weathering on a global scale and a geological cycle rather similar to that in action today could begin.

So far so good; the planet earth had taken shape and was spinning on its axis and revolving about the sun in orderly fashion. But on the surface of this new planet and within its interior many changes may have been taking place. The evolution of the solid planet was under way and in its youth it may have been unruly and restless, expending great energy, as is the case with other kinds of youthful bodies. This turmoil may have been expressed at the surface of the earth in the volcanic tirades and structural disturbances mentioned above, but here we have first to turn our attention to the origin and early history of the outer crust. To be truthful, we are in this book really concerned with little more than what happened to, and on, this outer crust in the later history of earth.

To sum up so far, then, no matter whether the protoplanet was originally hot, or cold, it would have become hot at the centre by gravitational attraction. Temperature at the centre would gradually rise to 1,000 degrees centigrade or more by this means, and with radioactive heat production at about five times its present level, the temperature would have soared to several thousand degrees higher. The American geophysicist Francis Birch has calculated that somewhere between 500 and 1,000 million years would be needed to heat up the interior of the earth to the point where nickel-iron melts. As the earth became molten the nickel-iron migrated inwards, gravitated, to form the core. The lighter materials, largely silicates, were left behind as outer layers, mantle and crust. All this probably took place 4,500 to 5,000 million years ago, judging from the oldest radiometric age determinations we can find.

The mantle would probably have become more or less solid before the outer core solidified. This may seem to be an odd situation but it is explicable as follows. The mantle consists of magnesium- and iron-rich silicate minerals which have relatively higher melting points than iron at any combination of pressure and temperature. Using data from electrical conductivity measurements, geophysicists in several countries have estimated that the temperature at the bottom of the mantle is between 4,000 degrees centigrade and 5,000 degrees centigrade. At that depth and temperature nickel-iron would be truly molten but silicate minerals of the mantle would begin to crystallise and solidify.

Meanwhile at the outer surface the crust may have only been partly melted. Some geologists do not believe that the entire earth was ever molten, because the heat needed would have driven off practically all the volatile elements. Zinc, arsenic, mercury and all the familiar gases would have been amongst these volatiles, but since they are present in considerable quantities perhaps the earth never was completely raised to a temperature where they would volatilise. Perhaps the radioactive heat of the earth was only enough to lower the viscosity of the interior to a point where slow convection of heat could take place and the elements could migrate to the appropriate levels in the earth by solid diffusion. For most of us these ideas are difficult to grasp, especially since we are dealing with temperatures and pressures well beyond those familiar to us in everyday life. Moreover we cannot get at the earth's core or the mantle, we can only sample the outer crust.

For many years now geologists have realised that the composition of the continental crust approximates to that of granite and that granite can be produced by the crystallisation and concentration of the lighter elements from a basaltic magma. The crystallised lighter material represents only about 7 or 8 per cent of the original magma, and of course it floats on the top apparently just as the sial floats on the sima. Some geologists feel that the sial or continental crust was formed at an early period in earth's history: others believe it has been formed progressively throughout geological time. There is enough basaltic material in the earth to give rise to the lighter rocks of the continents many times over. The sial is only about 0·005 per cent of the mantle in volume. More sialic crust could arise in the course of time to come. To account for the relatively small volume of sial present today Charles Darwin's astronomer son, George Darwin, put forward the suggestion that much of it has been torn away to form the moon, but the objections to this idea were numerous and compelling. The final nail in its coffin has been the collection of moon rock samples by the Apollo teams: the rocks are all basaltic.

The oldest continental granitic rocks appear to be about 3,900 million years old and we can assume that until that time the crust was formed, locally melted and assimilated into the sima and congealed again. All in all this would have been going on for about a thousand million years after the separation of the core and mantle. There was still prodigious volcanic activity and crustal turmoil but the point had been reached when there was no longer enough energy within the planet to melt all the granitic material at the surface. Large parts of the crystallised, frozen lighter material have persisted right to the present day. From time to time large masses of the crust have been 'cooked' up again. And although it has clearly changed its shape and position on the face of the earth from time to time, the continental granitic material has tended to remain in lumps, slabs or 'sialbergs', rather than spread out as a much thinner but uniform skin.

It is difficult to explain why this should be. Put in another way, it is the problem of the presence and nature of the ocean basins that is so perplexing. While George Darwin's idea was that the Pacific Ocean is the scar left after much of the sial had been flung or torn out to compose the moon, some geologists have favoured the view that the once-complete skin of sial has in places been assimilated, or dissolved away in the areas now occupied by the ocean basins. There is no real support for this hypothesis today; all the evidence seems to militate against such colossal destruction of the original 'skin'.

From what we shall see in Chapters V and VI the continents are far from being simple uniform slabs of granite. They possess complex internal anatomies, acquired or evolved over the last 3,500 million years or more. Each continental mass is the site of the rock cycle in action, today as much as in the past. The forces that produce orogeny have tended to locate the most active segments of the cycle near the margins of the continents during this last 3,500 million years, and the continents may have grown rather than diminished in area as a result of this. The original sialic differentiate may have appeared at the surface where those forces were primarily most in evidence. The motive power behind orogeny and volcanic activity must be terrestrial heat and the movements it causes in the mantle. From time to time in the preceding pages we have mentioned convection in the mantle as the possible manifestation of the flow of heat out from the deeper parts of the earth. At the points where both the upwelling and the down-drag of the mantle took place there would have been physical and chemical activity enough to allow the differentiation of the sialic fraction. Perhaps during the course of time the convective movements have persisted in such a way as to allow the 'sialbergs' or rafts to break up and to rejoin from time to time. But because there has never been a complete fusion of the crust there has never been a universal melting of the sial to allow it to spread in all directions. Quite the reverse seems to have taken place, but we anticipate the contents of our later chapters.

Moon-rise

Meanwhile there remain the questions of the origin and evolution of earth's large satellite, the moon. Geologists have until recently not paid to the moon the attention it has deserved, but the Apollo programme may have far-reaching effects upon our concept of the earth–moon partnership. Our satellite has importance for geology in two respects: it influences the earth's motions in space and it is perhaps a dead or 'fossil' planet which has had little or no change in its surface since its formation. Despite the differences between the earth and the moon, the origin of so large a satellite must be related in some way to the birth or early history of the earth. Both the rotation of the earth about its axis and its movement about its centre of mass are affected by our single moon. The presence of the moon influences the length of day, and the moon has always provided man with a form of calendar.

The moon is about a quarter of the size (radius 1,738 kilometres) of the earth which it orbits in an ellipse inclined at about 5 degrees 9 minutes to the earth's plane of rotation about the sun. We believe that the eccentricity of the moon's orbit is very slowly increasing. It takes 27 days 7 hours 43 minutes and 11·5 seconds to complete a revolution, the sidereal month. During this period the mean of its distance from earth is 384,000 kilometres, less than 1 per cent of the distances between earth and its nearest planetary neighbours—Venus or Mars—at their closest.

Not only is the moon smaller than the earth, its density is only 3·34, compared to earth's 5·52. To account for this, an overall composition approaching that of the upper part of the mantle of the earth may be imagined. One result of the comparatively small mass and density is that the force of lunar gravity is only about a sixth of that of the earth, as the moon-landing parties have found to their amusement and concern. There is no lunar atmosphere nor hydrosphere and the lunar temperature swings from 130 degrees centigrade at 'midday' to −200 degrees centigrade at 'midnight'.

The topography and geology of the moon's surface is becoming known in detail, and explanations for the lunar craters, marias, rilles and 'stratigraphy' are currently much debated. Lunar samples—basaltic rocks with a variety of ferromagnesian minerals—have been analysed and provide a large body of lunar chemical information. There are some immediately obvious differences in composition between the earth and the moon. The ratio of iron to nickel is larger; nickel, sodium and potassium are much less abundant while carbon and nitrogen have so far been detected only as minute traces.

Analyses of lunar structures and material are important in providing the basis for a discussion of the origin of the moon and its relation to the earth throughout geological history. Moon rocks give age determinations of up to 3,500 and 4,600 million years, and the formation of the moon is believed to be of the same date as that of the earth. There seems to have been no universal surface melting since 4,600 million years ago—the age of the moon 'dust'.

The moon's craters may have been produced by meteors crashing into its surface and the great marias or 'seas' are perhaps frozen floods or huge lakes of lavas formed as the result of a rain of meteoric impacts. An age of 3,700 million years has been found from the analysis of lavas from the Mare Tranquillitatis and this may have been a time when not only the moon but the earth as well suffered a celestial bombardment.

So how are we to account for the origin of this old, barren, scarred satellite of the earth? George Darwin, whom we have already mentioned, pointed out that the angular momentum lost by the earth cannot just disappear. It is now calculated that the momentum is transferred to the moon to the extent of moving the moon away from earth at a rate of 5 metres per century. This is not a great amount in itself, but it could have a real significance if continued for a long time. If the rate has been constant over the thousands of millions of years that the moon-earth system is thought to have endured, it follows that earth and moon must once have been very much closer together. George Darwin used this idea to propose that the moon had in fact been born by fission from the earth, tearing away from where now the Pacific basin is found. Most geologists have treated the hypothesis with scepticism, feeling that such a parting would be too violent to allow the remaining part of the earth to survive in one piece. Samples of rock from the floor of the Pacific are no older than 120 million years but samples from the moon are very much older. Moreover, we know today that the Pacific Ocean floor is composed of at least three of the major plates that make up the crust of the earth. No origin for the moon is possible there.

Here another ingenious suggestion has been made to account for the moon—a suggestion to be tested by the Apollo programme's results. When the earth was still very young and perhaps not yet clad in a solid crust or differentiated into iron core and outer shells it may have been a rather flatter, disc-like liquid body. It was also spinning much faster with a day lasting about four hours. If the skin ruptured, blobs of the disc might be flung out like mud from a fast-spinning car wheel. Once shot out from earth, these blobs could either go into orbit around the sun or fall back to the earth. The problem has been to provide force enough to lift the mass of the moon from the earth to give it an orbit around the parent planet. It is solved if we imagine that not only the moon but Mars also was spun off the rapidly rotating earth. Mars escaped from the earth's gravitational field only to be captured by the sun's field. The size and density of Mars is not inconsistent with this. The moon, however, was the unfortunate small droplet that was left behind at a distance near enough to earth to go into orbit but not too near to be pulled back all the way. It is a mere eightieth of the mass of the earth. From what we know so far the moon also seems to lack an iron core. In spite of the attractions of this hypothesis for the origin of the moon many scientists are critical of it.

One astronomer has thought that the moon was perhaps captured by earth from an original home in the asteroid belt. To have done so without involving a colossal bump, with all the attendant catastrophic side effects, seems unlikely. Another suggestion has been that the moon gradually formed from a dust belt or ring around the earth, or possibly the earth and moon formed simultaneously but quite independently from a massive dust cloud. A nearby moon would mean greatly increased tides and there is plenty of evidence of strong tidal activity in the deposition of sediments throughout the geological record far back into Precambrian time.

Tides are obvious in the oceans of the world but the effects of the moon's pull upon continents are not so easy to determine. A land tide of 20 centimetres (8 inches) has been estimated, and has long been held to account for the systems of vertical partings known as joints, which occur in most rocks, and even for some faults.

We have known for about 250 years that the moon appears to be speeding up in its orbit round the earth. Edward Halley, the discoverer of the famous comet, first demonstrated this, and even earlier the German philosopher Emmanuel Kant had suggested that the ocean tides had a braking effect upon the earth and were slowing down its rotation.

A new idea concerning the influence of the moon during its existence has been advanced by British geophysicist S. K. Runcorn, using the information from what may seem to be an unlikely quarter. Devonian corals, dated as about 370 million years old, are now known to have secreted skeletons of calcium carbonate, calcite, in a very regular way, adding tiny rings of it to the top of their skeletal cup as they grew. This they did day by day, and we find that they possess daily increments of calcite in regular units of 400 rather than 365. At that time the day would have been 21·9 hours long. There is also a cyclic arrangement averaging 30·6 growth ridges, representing monthly additions. These two values help to assess the average rate at which the tidal-braking action has operated, and the result coincides with that obtained by the astronomers. What this implies is that the earth and moon were in contact not 4,600 million years ago so much as only 2,000 million years back. But this is clearly impossible from what we know of the age of the moon rocks and surface features. The moon must be as old as the earth, and no matter how changes of sea level affecting the efficiency of the tidal brake are viewed, the puzzle of the moon's origin is not yet solved beyond doubt.

All through the debate on the age and evolution of the earth geologists have tended to adhere to one of two schools of belief: either that the earth is slowly shrinking by cooling or that it is expanding. The latter school maintains that the expansion is causing ocean-floor spreading, but the idea has no great following amongst geologists at large. Many of the most active students of earth history and structure hold that mountain-building movements and ocean floor spreading are dependant upon great convective flow systems within the mantle. One might ask of these geologists if the development of these convective cells could mean that the heavier elements in the earth have been gravitating towards the centre throughout time, thereby causing the earth to spin faster. The evidence of the corals hints that a rate, if not a mechanism, is at hand.

So far no trace whatever of organic activity or material has been found in lunar samples. In the absence of an atmosphere and a hydrosphere, forms of life comparable to those on earth are not expected, but the sterile nature of the examined moon rocks suggests that life never was present. Had there ever been an atmosphere and life on the moon's surface before the tremendous bombardment that produced the lunar craters it would, no doubt, have left some traces in the rocks. The N.A.S.A. geochemists are not inclined to be very hopeful of finding such traces, but they are nevertheless methodically analysing every kind of material that is available.

The debate about the origin of the earth and the solar system continues with renewed vigour as space research, especially by the U.S.A. and U.S.S.R., continues to bring in new and apparently sometimes contradictory facts. For the moment earth scientists have enough of a picture of the events before geological time began that is consistent with the principal facts to keep them reasonably happy. This is not to say that they will not continue to argue about it with relish, but the better the arguments the happier the scientific community at large tends to be. They can feel that the stage has been set, although perhaps rather sparsely, for the next act, the eons when geological processes were to produce a record that is still in part available.

CHAPTER V

What Goes Up Must Come Down:
Mountain Ranges

IT MAY seem an odd turn of the argument that takes us from a view of the earliest crust of the earth and the hot and hellish events of 3,500 and more million years ago to the ethereal cool of today's mountain ranges, but there is a reason for it. Considered from almost any standpoint, mountains are remarkable features. By definition a mountain may be any land mass that projects conspicuously above its surroundings, and by common consent those parts of the continents that rise above 1,000 metres or so are regarded as the mountainous regions. Almost all of them, as we have seen, run in long belts or cordilleras, some narrow, some wide. In total they make up about a quarter of the land surface, which is not an inconsiderable portion of the continental area. Hardly surprising, then, is the fact that mountains influence climate, affect life and exercise a remarkable control upon the nature of each continent. They hinder and divert the orderly flow of the air streams and winds about the earth and they influence the temperature and rainfall of wide areas not only nearby but also stretching for thousands of miles.

The influence of mountains upon the history of mankind has been important and they exercise a fascination that reaches almost everyone. They have featured in the mythologies, folklore and religions of most peoples. Perhaps it is a truism to say that much of their attraction for civilised man lies in the strangeness of the relief and the changeable character of their climate. This is no place to develop an essay on why mountains attract men. There are libraries devoted to the topic. But it does no harm to recall that mountains not only have their own physical characteristics and special living inhabitants but also that they have always influenced man's thinking about the nature of his world.

For most geologists they are manifestations of some of the greatest forces within the earth. The might of the process known as orogeny, the strength that has pushed up mountains ever since the crust was formed, is clearly part of the primeval energy that keeps our planet evolving.

Mountains have acquired the reputation of permanence, the property of always being there, immutable and immovable. 'As old as the hills' signifies great age and, presumably, stability. For the geologist such a comparison will carry little weight, no guarantee of an ancient pedigree. The majority of mountains today are, geologically speaking, not all that ancient; and the higher the mountain, the younger it may be. The great mountain areas of the Alps, the ranges of the Himalayas and the Western Cordilleras of the Americas are much younger than most of the continents that border them. Yet in many respects they are the most significant parts of the continents and they reveal the effects of some of the colossal forces that have been operating throughout earth's history. It has been for these reasons that mountains have exercised such an attraction for geologists. Here we can see that there is a kind of rhythm and pattern to the way in which strains within the crust of the planet have been relieved by the folding, crushing, melting and wholesale transport of surprisingly large volumes of rock. Here too there may be readable signs that these forces are still at work. The sources of such energy and the means by which they produce the wrinkles we call mountains are major concerns for the geologist.

Large and high though the mountain areas be, they are really quite small relative to the size of the planet itself, and perhaps small compared to some of the mountains present on the moon or on other planets. It has been said that a large mountain may be proportional in size to a pimple on the back of a human neck—and in proportion about as long-lived. Every continent has its mountains, and seemingly always has had. To be sure, as the ages have passed the sizes, shapes and positions of mountains have changed. Like men, they are born, grow, decline and return to earth, and like men they leave a testimony of their existence. As we shall see in the next chapter, the evolution and destruction of mountains seems to be an integral part of the growth and evolution of continents. Of course all this requires time, and our means of finding out how much time are still rather crude.

There are, naturally, several different kinds of mountains. The individual peaks may share only some of their history with others in the same range; adjacent ranges may differ in age and structure. Many mountains are volcanic, isolated, others are composed of rocks that were once sediments buried deep on the sea floor. The ancient Greeks were familiar with sea shells embedded in rocks now high on the mountains around the eastern Mediterranean.

Following the Renaissance and during the eighteenth century men began to enquire more energetically into the nature of mountains,

Leonardo da Vinci included quite spectacular mountains and crags in the backgrounds of some of his paintings. His notebooks contain sketches of strata and contorted rocks which he must have carefully observed in the highland regions of Italy and France. Long before then, however, mountains were known to be locally rich in minerals, precious stones and splendid building and ornamental materials. Then again, were not the mountains the sources of the great rivers of the world?

That remarkable French aristocrat, the Comte de Buffon, some 200 years ago, commented on the origins of marine invertebrate fossils in the hills of France. And in Scotland, as we know, James Hutton was so impressed with the relationships of strata on the coastline that he was led to produce his *Theory of the Earth* which in its way was as revolutionary as any product of the 'Age of Reason'. Hutton recognised that mountains are geologically ephemeral, not lasting for ever, and he recognised that they are perhaps the regions in which the strongest geological forces of construction are matched by the most vigorous destructive agencies. Rain, snow, frost and sun, the extremes of heat and cold coupled with all-pervading moisture and all-important gravity, never cease to wear away the rocks thrust up from below. Not only is this so now, but as Hutton saw it, there was no vestige of a beginning nor prospect of an end to this cycle.

Buffon was nothing if not prodigious in his interests and prolific in his writings. His 35-volume work, *Histoire naturelle, generale et particulière*, was an attempt to record all that was known of the world of nature, and inevitably he had much to say about the origin of such big geological entities as mountains. The eighteenth century was a time when some rather fundamental questions were being asked in France, about nature and the scriptures as well as about politics and social matters. Buffon's proposals that the biblical account might not be adequate to explain the history of the world were in keeping with the intellectual curiosity that was growing in Europe. In other countries as well as France his works were well received. They presented an orderly account of an orderly sequence of events in earth history—seven epochs, occupying 75,000 years or so, during which earth evolved from a hot and molten state to its present condition. In other European countries geological columns and divisions of the crust were being recognised. At Freiburg Mining Academy in the latter half of the eighteenth century the great teacher, Abraham Gottlob Werner, was demonstrating his ideas on the Neptunian origin of the crust, and of granite in particular. The conflict of the Neptunists and the Plutonists has drawn our attention previously, but in one thing at least these two bitterly opposed schools of thought were agreed. Granite is commonly at the core of mountain ranges, a universal foundation for all later rocks and structures. Granite was heaved up when the mountains took shape.

It was indeed the heaving-up that impressed so many of those interested in the history of the earth in Werner's and Hutton's day. Such violence, such changes were envisaged, and so drastic an effect would these have had upon all life, ran the thoughts of these worthies, that it must have been accomplished in a universal catastrophe. Thus the catastrophists found an explanation for the contortions and displacements of the rocks. As we know, they were routed by Charles Lyell's masterly demonstrations of Hutton's ideas and his thesis that in the past earth movements were probably no more intensive than they are today. Undeniably, there seem to be cycles of events in the rise and decay of geological features, the advance and retreat of the seas over the lands and in the resurgence of volcanic activity. Lyell concluded that cycles of a sort had been operating throughout geological time. Our modern view is that perhaps the orderly chains of events pictured by Lyell may only have occupied a small part of geological time and that the earth is evolving by an expenditure of energy from both within and without.

It was only natural that the fiercest arguments of the catastrophists and the uniformitarianists should centre on the more complicated geological features. The Alps and the worn-down remains of more ancient mountains in western Europe were the regions in which they found the fuel for their disputes. The Alps in particular afforded a splendid battleground; there are sedimentary, igneous and metamorphic rocks close together and the field relationships of these rocks can be seen in the mountainsides and gorges. This geologically complex region with its majestic peaks and great ranges is typical of the mountain chains so conspicuous on the globe today and it belongs to the Alpine-Himalayan 'mobile belt', a region where all kinds of geological movements have been active. However, before we explore mobile belts further we should briefly look at the broadly different kinds of mountains that occur not only in the mobile belts but also in regions which have long since ceased to be mobile.

Going Up

Mountains can grow in four different ways or a combination of them—or to be more accurate, there are four ways in which the crust can be built up high above the general level of the continents. It is the drastic erosion and the scouring which these uplifted regions suffer that produces the great relief and jagged topography of mountains. But the uplift is necessary first. By far the greatest number of mountain ranges has been produced by a pushing and compressing of part of the earth's surface. This is rather analogous to ruckling a table cloth spread on a table when the table leaf extensions are withdrawn. The folded arches and domes (anticline) or rock may topple over and continue to deform as overfolds or *nappes*. The mobile belts with their folded rocks are clearly full of examples of this mechanism and we must return to them a few pages on.

Another way in which large volumes of rock are displaced and elevated is by the action of faults. Under strong compression portions of the surface rocks may fracture rather than bend and breaks or faults are produced. Such faults are generally clean breaks, anything from a few metres to several hundred kilometres in length, and the rocks on one side are moved relative to those on the other. When the movement takes place there may be violent earthquakes. Strata or other rocks may be displaced by several metres in a severe quake. Landslides and mudflows may follow. It is common for faults to occur not singly, but in numbers, more or less aligned in the same directions. Although most faults are steeply dipping planes or zones, with one block moving predominantly up or down relative to its fellow, the movement may be lateral or transverse. Here the blocks don't change their level so much as their horizontal location with one grinding or gliding past its fellow. The great San Andreas fault of California is apparently such a 'transverse fault', but, as is usual, there is some vertical displacement along it.

In most of these cases the plane of the fault itself approaches the vertical, but some of the most spectacular fault movements are along planes that are close to horizontal or perhaps undulating. These low-angled affairs are called thrusts and there may be a displacement of the upper block or slab to the tune of scores of kilometres. Their effects can be seen in the Alps, the Appalachians and the Rockies where whole ranges have been pushed along in this fashion. Thrusts are identified in most of the mobile belts and in almost every case are parallel to the length of the mobile belt in which they occur.

Two or three decades ago there was an interesting controversy about the way in which movement had taken place along certain of these thrust planes. Quite clearly the planes themselves dipped down in the direction of the movement instead of in the usual opposite way.

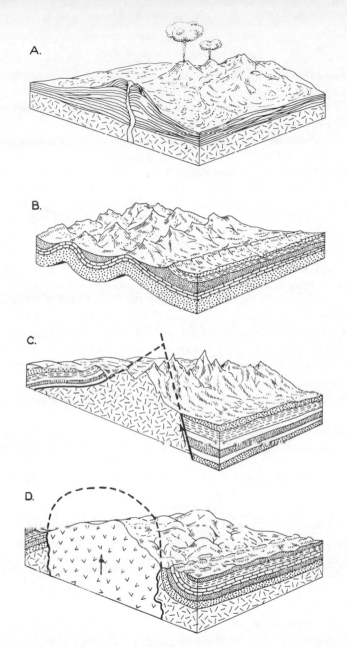

Figure 5-1 Four basic types of mountains can be distinguished; A, those that are volcanic cones; B, those that consist of folded strata; C, upthrust portions of the rigid continental basement; D, dome-shaped bulges produced by the intrusion of masses of igneous rock at depth. Many mountain ranges are made up of several of these types and some individual mountains combine the features of two or more of these basic types.

Here the thrust block was not so much being squeezed up out of the earth as sliding down a gently inclined plane from some higher region. Under the influence of gravity, masses of elevated surface rocks were slipping off an uplifted core. The thrusts develop where the rocks are easily ruptured and give way laterally; shales or clays may provide these weak zones along which slipping develops on a grand scale.

This interpretation of movement in a mountain range is not so very different from that suggested by Hutton two hundred years ago. He pictured an upwelling of granite causing overlying schists, slates and other rocks to slide away, folding as they went so that thrusts and overfolds of strata were produced on the flanks of the granite.

Granite is a rock to which our discussion returns time and time again and it is the rock frequently associated with the next kind of mountain-forming activity we shall examine. Elsewhere, basalts are the rocks essentially involved in raising the level of the ground. The results of this movement are usually called 'dome' mountains and they result from a welling-up of molten rock from below to accumulate like a gigantic blister just below the surface. In short, we have a bulge, a bulge that may be several hundred metres thick. The name 'laccolith' is given to this kind of intrusion of igneous rock. Some of the largest of these intrusions gave rise to the Henry Mountains, Utah, in the U.S.A.

Much larger bulges found in the very heart of mobile belts, both ancient and modern, have been produced by a pushing up of granite. These deep-seated masses, known appropriately as *bathyliths* (deep rocks), seem to have formed in the hottest part of the mountain belt, several kilometres below the surface. Under great pressures from the weight of rocks above and the high temperatures below, the granite may corrode and force its way upward along any available line or passage of least resistance. Before it can break out at the surface, however, it congeals and crystallises at depth. No one has ever seen the bottom of a bathylith. The granite or other intruded rock is to all intents and purposes a squeezed-up blob or bulge of the granite layer beneath the continents. From the way in which the granite corrodes and eats into existing formations we judge that bathyliths form only when intense pressures have already greatly deformed the local rocks.

Huge areas of granite bathylith mountains occur in the Coast Ranges of western Canada, in California, and in the Andes of South America.

Finally the fireworks, the mountains formed by volcanic activity. It might truthfully be said that the greatest mountains of all are essentially volcanic in origin—the mountains that rise from the ocean floors to dot the oceans with basaltic islands. Many, but not all, of these 'mountains' are part of the oceanic ridge system and are perhaps some of the most fundamental features of the surface of the earth. Only when they emerge from the waves can we see how this volcanic activity takes place and how it can build up immense piles of rock matter.

In Chapter II we noted that for many years the distribution of volcanoes has puzzled geologists, but that we now seem to be nearer to understanding it than previously. Why should conditions prevail only in these relatively restricted areas wherein molten rock, magma, forces its way up from the deep hot regions beneath the surface? And what is the significance of the distribution of different kinds of volcano, of basaltic or andesitic lava?

It is now generally agreed that magma reaches the surface from great depths through pipe-like conduits or through long fissures that extend down to the hot region of the upper mantle. Just how many kilometres down this is may vary from place to place. Something is needed to trigger off the escape of magma to the surface. Basaltic magma seems to be relatively near the surface in the vicinity of the mid-oceanic ridges. Elsewhere, however, the magma may originate at much greater depths near the edges of the continents. The Benioff zone has already figured on these pages as the region where deep-seated earthquakes relieve the great stresses beneath the crust. Earthquakes involve friction and the generation of heat, and in the sloping Benioff zones at the edges of the Pacific there seems to be a situation in which enough heat could be accumulated to melt portions of the crust and cause eruptions via some of the local fractures.

Geosynclines Forever!
The Alps, the Himalayas, Rockies and Andes are all to a very large degree fold-mountains. The strata within them have been compressed and concertina'ed into overfolds and nappes, faulted and thrust, intruded by granites and generally given a very rough time since first they were deposited. Along the borders of these great ranges faults and thrusts abound. Overthrust blocks may lie several deep and above and behind them there may be giant folds, overfolds and arches. The ground against or over which these rocks have been pushed is, however, relatively little disturbed. Deeper within the mountain lands the folded structures may give way to relatively less deformed plateaux and highlands, or there may be a bathylith or other igneous

mass. In traversing some of the mountainous regions of the world the early geologists felt that there was a sort of symmetry to the geology of these regions, as though the strata had been crushed between the jaws of a giant vice. The rocks folded, broke, flopped over outwards onto each jaw of the vice as the pressure was maintained. The analogy broke down in many cases as exploration increased but in every instance there seemed to be an intense squeezing up of rocks so that quite ancient rock formations spilled over onto younger and less deformed units on one side or another. In the eyes of many geologists in the first quarter of this century, the compression could be produced by a contraction and drawing down of the crust into a long trough. Nearly a hundred years ago the Austrian geologist Leopold Kober visualised the Dinaric Alps and the Carpathian Mountains of Europe as the thrust and folded mountains bordering a central relatively undisturbed mass, the Plains of Hungary. The vice here was squeezing a tract of country about 1,000 kilometres wide.

Apart from being impressed by the enormous squeezing and alteration that many of the rocks showed in the great mountains, the geologists of the mid and late nineteenth century were struck by the fact that these mountains consisted of marine sedimentary rocks that, layer upon layer, were more than 10,000 metres thick. It was noticed in the Appalachian Mountains that something like this thickness was involved but that out in the Mississippi Valley rocks of the same age were only a small fraction of this thickness. In the Rockies and in the Himalayas thicknesses at least as great were evident. And the larger part by far is made up of rocks formed in relatively shallow water.

Tackling the problem of how such a great accumulation could occur in a region that was subsequently to become mountainous, the great early American geologist, James Hall of New York, laid the foundation of one of the major concepts in geology. Hall was in 1857 president of the American Association for the Advancement of Science. This was appropriate because the A.A.A.S. had in fact begun as a geological society. Hall pointed out the great thickness of the Palaeozoic rocks in the Appalachian Mountains compared with the Mississippi Valley and offered the explanation that as the weight of sediment accumulated it had pushed down the crust beneath it. He believed that eventually the crust could take the strain no longer. It buckled and the strata were crumpled and raised high above their original level.

A different explanation was forthcoming in 1873 when another American geologist, the renowned James D. Dana, rejected Hall's

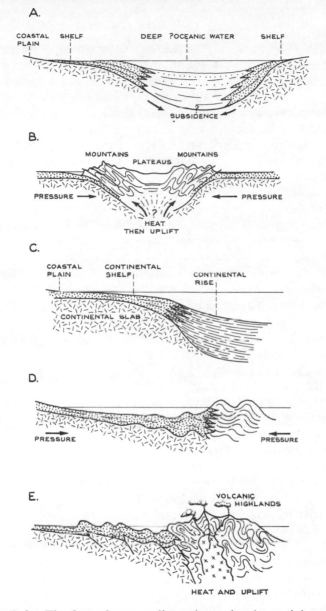

Figure 5-2 The fate of geosynclines always has been violent. In A we have the original view of a geosyncline and in B the orogen produced by squeezing the geosyncline is a nicely symmetrical affair. Unfortunately few orogens are known to be like this. C, D and E give the modern view of how thick geosynclinal formations (many with volcanic rocks not shown here) may be deposited and then squeezed at the edge of a continent.

explanation of subsidence by loading the crust. He offered a new interpretation, namely that the down-warping of the crust was a cause not a result of the thick column of sediment. A long deep depression in the crust offered a site for the accumulation of sands, silts and other sediments over a long period. This phase of down-warping and sedimentation gave way to one of uplift and compression. The trough was referred to by Dana as a *geosyncline* and the association of geosynclines with mountain building has now been demonstrated in many parts of the world. Hall and Dana were lucky in that the Appalachian geology is relatively simple. When European geologists studied their own mountain areas they found that, despite local complications, the geosynclinal concept could be applied to explain the features present.

The early part of the twentieth century found geologists in many parts of the world seeking to explain folded mountain ranges and their history in terms of geosynclines. During this while the realisation was borne in upon the investigators that geosynclines are not at all simple affairs. The Appalachian Mountains south of New York City are relatively unrepresentative in this respect. The development of a large trough-like depression in the crust takes millions of years, it proceeds now by fits and starts, now by slow continuous down-warping, while sediment is washed in from land elsewhere.

The weight of sediment several kilometres thick is great indeed, but it scarcely seems competent to push down the floor of the trough enough to hold the masses involved. And, anyway, where did so much sediment come from? There must have been a corresponding uplift of land on one side or both sides of the geosynclines to provide all the detritus. Not always can we find traces of where the land lay nor how it was behaving, but it would seem to have been a very deeply eroded and planed off land mass by the time so much of it had been through the erosion mill.

As the geology of the various geosynclinal areas has been unravelled it has become apparent that the crust within the trough was in movement almost continuously, sometimes it rebounded upwards here and there, elsewhere or at other times it sagged and flexed. Another aspect of 'geosynclinal geology' that has to be noted is that all the strata, now folded and faulted so intensely, would occupy a very wide area indeed if they were to be returned to the kind of attitude that they must originally have had on the sea floor. As a result of the unrest on the geosynclinal sea bottom and the wide areas perhaps involved, it is small wonder that the strata themselves are so varied and difficult to relate to one another from one place to another. Many of the sedimentary rocks are dull, compact muddy sandstones with few fossils, quite unlike the very fossiliferous limestones found adjacent to many old land masses and on continental shelves.

In the course of tens of millions of years the oldest sediments in the geosynclines were forced down by the accumulating material to levels where they were surrounded by denser rock. From here on they would tend to float, their continued presence so deep down could only persist if the weight above remained or if pressure from the sides kept them in place. What subsequently happened to the sedimentary layers on the floor of the geosyncline seems clearly to have been related to the behaviour of the crust immediately below and flanking them. There are not many areas where we can see what was actually underneath the old geosynclinal floors, but in some instances it seems to have been basaltic material, like oceanic floor, or even denser rock, resembling perhaps mantle material. On each side, however, or on one side, there was rather rigid continental crust. Geosynclines always seem to have formed at the flanks of continents, either in the space between two continents or along the oceanic margin of one.

No matter if flanked by continents or supported only by one, all geosynclines seem ultimately to have incurred the wrath of the forces within the earth and to have been squeezed, deformed and crumpled into narrow, elongated, granite-ridden tracts, uplifted in a short but vigorous upheaval into mountain country. By the time these compressive movements begin to take effect the pressure and heat acting on the lowest layers within the geosyncline reduce the rock to a very plastic condition. No longer is this the kind of material we are familiar with from shallow depths. In the course of the changes that take place, the volume of the original sedimentary rock diminishes, the mineral content is reorganised. Slow flow and contortion result from the pressure, and during this process igneous activity usually begins to make itself felt. Even during the early stages in the life of a geosyncline there may be volcanic activity, but later on there rise large bodies of molten rock generated at lower, hotter, levels. Just how these magma bodies are formed is not always clear and many an argument has taken place about the granites, granodiorites and other igneous rocks that have been found deep among the roots of old geosynclinal accumulations.

Thus, in what we might call its death throes the geosyncline yields to compression and becomes a mobile belt; it suffers the mountain-building movements that are so characteristic. Where the waves once rolled by, mountains now rise. On geosynclines are founded mobile belts. But this is rather putting the cart before the horse since a geosyncline can only form where the crust is mobile or flexible enough to provide the basin or trough to receive the sediment. Some of our modern island arc trenches appear to fill the bill as mobile belts but as yet they have little accumulation of strata. Give them a few score million years and they may return a respectable geosynclinal assemblage of rocks.

At the risk of repetition, it is worth emphasising that from the study of mountain chains in many parts of the world and of different ages there emerges a constant theme in the sequence of events in an orogeny. Even during the long phase of overall subsidence in a geosyncline there may locally be short spasms of disturbance when the underlying rocks are compressed or shifted. However, when the compression of the belt begins in earnest large almost horizontal thrust planes develop. They extend along much of the length of the old depositional trough whenever the squeezing is strong enough. Some thrusts run for hundreds of kilometres, moving over them may be slabs or sheets of rock hundreds of metres thick and scores of kilometres wide. Later on these thrusts may be folded and thrust again.

After the thrusting is under way, and for a while after it has ceased, folding of the geosynclinal strata may continue. The crests and troughs of the folds usually run parallel to the thrusts, at right angles to the direction of greatest squeezing. As it proceeds the compression reduces the width of the belt while the length remains unaltered. The sequence of thrusting and folding may be repeated several times, each occasion narrowing the width of the disturbed zone. With each great compression there is, down below, regional metamorphism of the original sedimentary strata. Most mountain belts undergo several phases of such metamorphism and metamorphosis. The type and kinds of changes brought about in the rock depend upon the pressures and temperatures generated deep in the crust and the rate at which the sides of the orogen seem to approach one another.

This is the time when the great deep intrusions rise from below and as the lateral movements of crust die away so the whole region begins to rise. It is isostatic movement and it goes on until the new, thick, squeezed pile of rock is in equilibrium, balanced, with the surrounding region.

There must be something rather fundamental to earth evolution in the mechanism of geosynclinal-orogenic activity. We have the remains of scores of geosynclines preserved to a greater or lesser extent in the rocks of our continents today. Geosynclines and mobile belts seem to have been around ever since geological conditions, as distinct from the torrid pre-geological state of affairs, came into being. Were we to straighten out some of the folds in say the Alps or the Rockies we would have to push the rocks back scores, even hundreds, of kilometres. Could the compression be due to crustal shortening and the contraction of the earth? Have convecting cells in the mantle been responsible? To answer these questions is not easy and for the moment is postponed while we deal with the questions of how the

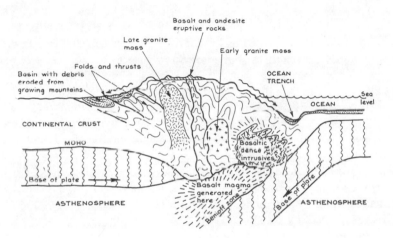

Figure 5–3   The big crunch comes to a mobile belt when it suffers such squeezing and compaction that there arises a highly deformed and internally heated ridge or pile along its landward side. In this way mountain ranges are produced

forces at the surface work to remove the upstart prominences of the orogenic or mobile belts and how they attempt to plane everything down to sea level.

Coming Down
'The higher up the mountain, the greener grows the grass' runs the old couplet, but higher still the grass finds it hard to survive. With altitude the cold increases, there is less moisture in the air and because of that ultraviolet rays are filtered less from the sunshine. In fact there is a strong zonation of climates and environments with altitude on

mountains and the ravages of nature upon the rocks are more conspicuous the higher one goes. High in the mountains of the equatorial lands and lower towards the poles lies the snow line above which there is snow and ice for the greater part of the time. Where there is snow, there may be ice, and together these two exercise a very profound influence upon the sculpturing of the mountain landscape. It's an influence as strong as that of water at lesser heights. Snow, ice, water, all have tremendous powers to move rock fragments and particles downhill, and frost, ice, and water have great ability to break up the living rock to provide those particles.

Snowfalls in the mountains may be thick and late into the spring or summer. On occasion the banks and fields of snow on steep mountain slopes may fall away suddenly in great avalanches. The destruction avalanches can cause is immense. Snow also feeds the glaciers, the rivers of ice that are born in the high corries and valleys and where snow and ice melt in the spring, raging torrents of water will spill down the mountainsides continuing the work of erosion and denudation.

One cannot fail to be impressed with the remarkable transformation of fluffy snow to hard, solid ice. The six-sided snow crystals are light lace-like affairs and trap a lot of air between them as they collect on the ground. As they accumulate and press upon the lowest snow crystals and with changes in the weather the air is pushed out, the delicate points are melted or evaporated away and the crystals reorganise as tiny rounded grains of snow, névée as they are called. More pressure, a little more melting, reduces névée to ice particles growing larger as time and conditions permit. At 25 metres or so beneath the surface solid ice may result.

Ice is one of the most remarkable of substances and as it accumulates in the high mountain regions it seems to flow out down the valley as a slow-moving mass, a glacier. Where the gradient is steep or broken the ice may break with crevasses, producing ice falls; where the slope is smooth it may seem to glide as a plastic mass. The rate at which flow takes place depends on the nature of the ground, the supply of snow and ice and its depth. Within the glacier the flow is far from uniform, being influenced by the nature of its channel, just like a river. The overall growth or decay of a glacier and its extent into lower ground is also variable and dependant upon snowfall and temperature. Some glaciers move several metres forward a day, others seem to be stationary and even in retreat back up the mountainside. During historic times the glaciers of the Alps and in Norway have been in constant fluctuation, now growing, now in retreat.

Glacial ice has immense power to erode the rock over which it moves and to transport the rock debris that is caught up within it or which falls upon its surface. It plucks and tugs at any projection beneath it. Where this loose rock material eventually reaches the sides or melting foot of the glacier it accumulates as piles of moraine. As a glacier dies from want of snow and its foot retreats from the lower ground the moraines remain like bull-dozed piles of rock and mud.

Impressive though moraines may be as the tell-tale remains of previous glacial activity, they are less spectacular by far than the land forms achieved by the carving and gouging activity of the ice itself.

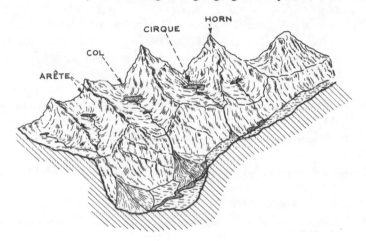

Figure 5–4   Glaciers in mountain areas scrape and gouge out major valleys, giving them the steepest straight sides shown above. They also sharpen and fret the watersheds and peaks.

The sand, fragments and boulders caught up or ripped off by the ice as it moves along add to its abrasive power and like a great rasp or plane it wears smooth the rock surfaces over which it passes. Quite highly polished and smoothed surfaces may result; some will bear striations, scratches and grooves which show the way the ice moved.

Cutting, grinding and scraping their way down from cirques to lowlands, the glaciers also tend to deepen their valleys by concentrating the most effective erosion beneath them. Instead of the characteristic V-shaped cross-profile created in valleys cut by streams, the glaciated valley has a U-shaped cross-profile, with a flat bottom and steep, even cliff-like sides. Small rock spurs are eroded away and the originally rather zig-zag course of a valley may become straightened. Tributary valleys without much ice may be left hanging high above the floor of the main valley. At the coast glaciated valleys may pass into fjords, deep, narrow arms of the sea. When sea level was lower, the fjords were in fact the lower ends of the glaciated valleys.

Meanwhile the spurs and ridges and watersheds between glacier-filled valleys are sharpened and accentuated by the shattering effects of frequent severe frosts. The snow and ice carry away the debris, and the smooth curves to the uplands that there once might have been are fretted and chiselled into sharp divides or arêtes. Several arêtes may join at a horn or Matterhorn peak.

During warm spells and in the summer there is a lot of free-running water on, in and under the glaciers. It removes rock debris, especially the fine rock flow ground by the sole of the ice mass, at a rapid pace. All this run-off may transport thousands of tons of material in a single glacier valley each year.

So characteristic are the erosional features produced by ice in mountains that it is locally surprising to find them also in areas well beyond the ice itself. Many are to be found in lowlands and mountains where there is no ice at all. The seemingly erratic behaviour of modern glaciers, sometimes advancing a few metres down their valleys for several years at a time and then melting away at an equally impressive rate for a while, was even more pronounced in the past. By keeping an eye on the weather and on the glaciers in the countries around the North Atlantic, Scandinavian scientists have found a relationship between snowfall, temperatures and winds and glacier activity. We know that in historic times quite drastic changes in the sizes of the glaciers and changes in the climate have gone hand in hand. To us, then, it is not too improbable a suggestion that at some time previously the glaciers extended very far beyond their present domains. But to many geologists of the nineteenth century the suggestion that ice had once covered much of Europe and North America was going too far.

Erratic blocks, carried more than 50 kilometres from their source in Switzerland, were known to James Hutton in 1795 but not until nearly 30 years later was the continent-covering scale of the Pleistocene glaciers realised. Appropriately this recognition was made in Switzerland by a civil engineer, Ignaz Venetz-Sitten in 1821. Within a few years the adherents to the biblical account of a great flood, Noah's deluge, being responsible for the many erratic blocks and the sands, gravels and clays spread beyond the glaciers were assailed by one of the most persuasive and literate of Victorian scientists. He was a Swiss, Louis Agassiz, later to take up an appointment at Harvard University, where he wrote a succession of papers outlining continental glaciation not only of Europe but of North America too.

The trail blazed by Agassiz has been followed and improved by generations of students of recent earth history. Their researches present us with a unique picture of a sequence of events that have modified the shapes and surfaces of every continent in the world and profoundly influenced the life of the last two or three million years. The Pleistocene epoch, the Great Ice Age, was in fact a time when the glaciers of the northern hemisphere grew to continental proportions not once but four times. Each occasion was one in which the mountainous regions, even in the equatorial latitudes, were under severe attack from snow and ice. During these cold times the actual volume of eroded material swept from the uplands to the lowlands, and from the continents to the seas, must have been very much greater than in warm periods of the same length. Between the frosty phases there were warm spells, though the amount of total precipitation in the mountains may not have altered all that much. Between them, ice and water continued hard at work, unceasingly and with increased vigour following every uplift of the mountains from below.

No less significant is the year-round work of streams flowing from ice-free mountains and from the lower reaches of those that have a mantle of ice. Streams in mountains tend to be unruly things, quickly rising with rainfall and moving quantities of soil, sand and pebbles as they do. Given a plentiful supply of moisture from mist, rain, and snow, etc., the mountains also have two other characteristics that render them susceptible to the erosive power of water. They are relatively free from a thick cover of plants that would serve to slow down the run-off, and they are high. All water runs downhill and the steeper the gradient the more is the power of water to carry away rock particles the stream may have.

Collecting into whatever depressions there are, water from rain, snow or ice soon begins to flow downhill and as it moves down a crevice, a fissure or channel, it does several jobs. It carries along mud, soil and rock particles; it erodes the channel deeper and locally where its path is arrested it may deposit sediments or deliver them to lakes or the oceans at the end of its course. Although sedimentary rock from the sea floor may be pushed up thousands of metres to form part of a high mountain range, weathering will disintegrate it; water and ice will sweep away the debris and the sediment returns once more to the bottom of the sea. There is no regular pace at which all this happens; it may involve many millions or even hundreds of millions of years, but it is a well recognised cycle.

Underlying the action of ice and of water is the ever-present force of gravity. It acts to move the rubbish and spoil of rock weathering and even unweathered bedrock to lower and lower levels. The slipping, creeping and falling of this material is known as mass movement or wastage and it is greatly assisted by moisture soaking into the ground. The water in the pores tends to push the individual grains apart and even to decrease the internal friction within, say, soil. So after heavy rains often there are mudflows, landslides and rock-falls.

On occasion these movements of the weathered rind of the earth can be rapid and extensive, not to say outright catastrophic. In mountainous regions most of them are small, scarcely to be noticed, but from time to time they are enormous affairs that do great damage. Spectacular rapid slides, falls or flows in inhabited mountain regions make headline news almost every year.

Mass movement at a very slow pace is harder to recognise and is less understood but it is nevertheless very important. Unceasing, but slow, mass movement down the upland slopes of the world probably accounts for moving much more material than do the flashy rapid events. Even with a cover of grass, this kind of destruction slowly lowers the land surface everywhere. Freezing and thawing help and where these are a common part of the scene the soil will flow down slopes in a process known as *solifluction*. The strange features produced in rock debris and soils by the frost are enormously varied.

The Life Span of a Mountain Chain
To nineteenth-century geologists studying the mountains of southern Europe and North America the presence of fossils in many of the strata high in the chains was a godsend. These fossils indicated the relative age or date of the rock. The stratigraphical time scale was broadly understood and the times when the Alps, the Appalachians or the Western Cordillera were pushed up could be estimated. By the end of the nineteenth century it was felt that geologists had a fairly good idea of the length of time the mountains had taken to form. Continuing side by side with these studies were those that concerned the regions made up of the remains of much older mountain structures. The stages in the life cycle of mountain chains had become known in

outline. The long period when sediments were accumulating, the phase of deformation, the phases of bathylithic intrusion, uplift and then vigorous erosion, were all recognised as being essential parts in this affair. Just how much time was occupied by each of these events in the life of a mountain chain was uncertain, but it seemed that the first stage, the accumulation of great thicknesses of sediment, took the greater part of the time. In contrast, the squeezing and uplift were very rapid violent affairs. Some geologists were so impressed by the speed at which vast upheavals took place that they called them 'earth-storms' or 'revolutions'.

When radiometric dating techniques came to the aid of the structural geologists it could be seen that the evolution of some mountain belts from geosynclines occupied about 200 million years, with the late compressive phases taking just a few million years near the end. However, by the time radiometric dating was being used it was also clear that some mountainous regions had histories that were not so simple, and which had taken a longer time. This was found to be especially the case with the older Palaeozoic mountains of Europe, and as the flow of age determinations continued it was recognised that the life of a Precambrian fold belt might have been very much longer. The activity of some of the Precambrian regimes might have extended to as much as 800 million years—longer than the entire Phanerozoic eon.

Estimates of the rate at which the actual uplift of mountains takes place suggest an average of about 70 metres per 1,000 years. This is about ten times as fast as the most active rivers could work to erode them down to a flat plain. To reduce ranges and cordilleras to sea level, the weathering and transporting processes have to keep hard at their task for a long time after the uplift has ceased. Between the local and relatively short-lived periods of rapid uplift, denudation has persistently gnawed away in its attempt to reduce the land all to a common low level once again. The present rate of erosion of the United States, where measurements have been made over many decades, would reduce an 'average' continent to sea level in 13 to 14 million years. But as erosion lowers the ground it also reduces the potential energy of rivers, and the actual time would be much longer. In fact the wholesale erosion of the land down to sea level is never attained. Mountains can, however, be reduced to low-lying plains in a matter of 10 million years or more. In an active mobile belt this is not a long period of time and its effects can soon be obscured by a few strong shoves from the orogenic forces within the crust.

Pause For Breath

So what went up as a mountain range, comes down particle by particle as sediment, debris, and in solution, carried by ice and water back to lower altitudes, back ultimately to the sea. The cycle of weathering, as it has been called, never ends. As soon as land emerges from the sea, the atmosphere and the weather, water and living things, attack it. Gravity pulls back the material that other earth-forces pushed out. There is a constant battle between the forces within the earth and those on its surface. Every moment material, somewhere or other, is on the move.

From time to time one or other set of forces prevails. When the mountain-building forces are in full play, vast areas of the crust may be raised thousands of metres in a short geological time. The agents of destruction attack the new uplands while they are still growing. Valleys, gorges and ravines are produced, narrow at first, widening later. As erosion proceeds, a great weight is transferred from one part of the crust to another, from the mountains or uplands to wherever the sediment is accumulating, usually the sea.

In the same way that a ship rises in the water as it is unloaded, so the lower parts of a mobile belt, the roots of the mountains rise under isostatic pressure as the mountain tops are worn away. As the ages roll by and the mountains of old are progressively reduced by erosion the more deeply buried structures are uncovered. The denudation gives us a view of the very foundations of the mountain chains of long ago. At these levels the rocks are intensely altered from their original nature. Everywhere are signs of heat and pressure and there are masses of granite and other igneous rocks. From the radio-isotopes in these igneous materials we know when they were emplaced and crystallised, and when the transformations at depth reached their climax.

In an earlier chapter we noted how a large part of the very foundations of the continents seems to be made up of old mountain roots, eroded and now out-cropping in the great shield regions, covered over by more recent, little-disturbed, sedimentary rocks in the great plainlands and pushed down from sight by the more recent mountains. Our next chapter will suggest how this patchwork of old mobile belts has originated, and how it seems to mark a fundamental process in the history of the earth.

CHAPTER VI

Relics of the Most Ancient Times

WHEN in the eighteenth century the first tables of strata were being drawn up by Abraham Lehmann and others in Europe, the orderly arrangement of formations with what we now call metamorphic and igneous rocks at the base was regarded as universal. Beneath the lowest strata are crystalline formations, the basement upon which all the later bedded units rest. These rocks must be older than the bedded formations which rest upon them. The question was 'How much older?' Geologists in Europe at that time thought of the crystalline rocks as having been precipitated from a primeval ocean. When Hutton and the Plutonists demolished this idea there were many problems still to be solved concerning the age and origins of these acknowledgedly ancient granites and other rocks. As geological knowledge grew, it seemed to indicate that the very foundations of the continents consist of granitic materials, woven into an impressive, baffling array of fabrics and structures. Where earth movements have warped, fractured and uplifted parts of the crust, these ancient rocks are now exposed at the surface. During mountain-building episodes segments of this basement have been pushed up beside, and over, much younger formations.

Nevertheless, the most striking and largest areas of basement rocks are the long-denuded and greatly up-warped, but otherwise stable, cores or heartlands of the continents, the so-called shields. They are distinctive regions, geologically as in other ways, and many of them have vast deposits of metallic ores and other industrially useful minerals. Then it is well to recall that the Phanerozoic rocks, representing only 15 per cent or so of geological time, cover about 80 per cent of all the land surface of the world. Shield rocks are not known in the deep ocean basins. The Precambrian is most widely exposed in mountain ranges and in the shields, elsewhere it may locally be revealed in crustal arches or warps. The geology of shield areas is sharply demarcated from that where other rocks crop out. To begin with, most of the rocks are, as the pioneer geologists pointed out, hard and crystalline and their fabric is deformed by the pressures they have endured in the crust. There are some rocks, it is true, that are not greatly altered, but they make up only a small part of the immense range of rock types in the basements. The shields are, for the most part, rather flat or only gently hilly country. Only in parts of the eastern Canadian shield and in Antarctica are there truly mountainous terrains composed entirely of Precambrian rocks.

At the margins of the shields and wherever younger rocks overlie them there are great unconformities, irregular surfaces cut in the basement rocks and smothered by the younger sedimentary beds. Only in a few places is there a conformable gradation up from the Precambrian rocks of the shield into later strata. In terms of earth history, the shields are composed of Cryptozoic rocks and the later formations are Phanerozoic. Most geologists use the term Precambrian (Pre-Cambrian) and some the Prepalaeozoic for the Cryptozoic, for that is what these rocks are, older than the Palaeozoic Cambrian system and, as the name indicates, containing mostly hidden or obscure traces of past life. In one or two places there are truly recognisable Precambrian fossils, but they represent only very humble forms of life and most of them are not easily identified. Radiometric dating suggests that the base of the Cambrian system has an age of about 600 million years. The very oldest Precambrian rocks known are about 3,900 million years old, but the majority were formed less than 3,600 million years ago.

The interval between 3,900 and 600 million years ago is a long one and we have no very detailed picture of the earth during this time. Basing our ideas on a knowledge of the Precambrian rocks of the shields, however, a few seemingly plausible generalisations can be made. In some ways, of course, the questions to be asked are more fundamental than those concerning the Phanerozoic rocks and times. After all, the 3,300 million years or so involved make up the larger part of geological time, and what happened then may well have laid down the pattern for events that have occurred since.

The Precambrian shields pose the problem of where this predominantly granitic material came from, and how the structures we observe came into being. In the previous chapter we asked if there has been a master plan by which the shields have formed in a single event or whether they have grown with the years. Do they suggest an expanding or a contracting earth? Are the continents being eroded away and their basement rocks ultimately to be dispersed as a layer of sediment spread across the ocean floors? To a few questions such as these, fairly definite answers can be given; the majority are not so easily answered.

The geologist concerned with Precambrian matters faces a problem that seldom bothers his colleagues dealing with later systems. There are virtually no fossils by which to correlate his rocks. Without knowing the relative positions and order of his rock units in the ground, the Precambrian geologist has some difficulty in establishing the kinds of structures present and in working out the geological history of his region. All has to be done on the basis of correlations made without fossils. The order of superposition and the cross-cutting relationships of intruded igneous rocks, coupled with the study of other physical and chemical characteristics of the rocks themselves, have formed the basis of most correlations and attempts to unravel the geological history of Precambrian formations. With the tool of radiometric dating in his hands, our Precambrian geologist is in a much better position to tackle the problems presented by these unfossiliferous and commonly much-deformed and altered materials.

During Phanerozoic times geosynclines and mobile belts appear to have evolved at the margins of the continents. They have left tracts of metamorphic and igneous terrain wrapped around or pinched between the relatively immobile or stable shields. It is not easy to distinguish a Phanerozoic crystalline rock from one formed much earlier in a Cryptozoic period. Radiometric dating has helped the structural geologist over this stile, enabling him to distinguish, for example, between the rocks that were metamorphosed in the Caledonian orogeny, say 350 million years ago in Scotland, from others nearby which were metamorphosed 1,200 million years or so back in the Cryptozoic era. For the chronicles of earth history this is an important matter.

The Cryptozoic Calendar

Before we can seriously attempt to understand the early evolution of the earth we need a means of identifying and of measuring the geological time that was involved. A geological clock is necessary, something that registers the passing of the years—only then can a calendar of events be drawn up, and tested against improved observations as science progresses. Radiometric analysis gives us the clock but the designation of the hours is a matter still being debated.

The earliest attempts at dividing Precambrian geological time were based on the simple idea that the oldest rocks were those that had suffered the greatest metamorphism. The logic of this is plain and undeniable, but it does not cover all contingencies. Under some circumstances older rocks may escape heavy metamorphism, while younger ones may be altered to an extreme degree. Notwithstanding this reservation, it became the practice to divide Precambrian time into an earlier period, the Archaean, and a later, the Proterozoic. It was during the Archaean that most of the coarsely crystalline layered rocks, the gneisses, were formed. For unspecified, but not unimaginable, reasons the metamorphism taking place in the Proterozoic was less intense; less altered rocks were produced. Indeed many Proterozoic rocks are scarcely metamorphosed at all. But once radiometric dating and detailed studies of the structural relationships of Archaean and Proterozoic rocks had begun it was apparent that the original grounds for distinguishing these periods were not valid.

From the Scandinavian geologists in the 1930s came a stream of investigations and reports that gave a new impetus to the study of the Precambrian. Their view of the Baltic shield was that it consisted of the worn-down stumps of several Precambrian mountain chains. The intense metamorphism had taken place deep down beneath the heaped-up pile of sedimentary and igneous rocks that constituted these mountains and its effects were now revealed in the eroded roots of the long-vanquished chains. The shield might be regarded as a bundle of such mountain root structures, and there are several groups of these, each distinct in area and in the way in which the structures are aligned. Each group could be regarded as distinct in age, one following another. Canadian geologists seized upon these ideas and pointed out that their shield, very much larger in area than that in Europe, contained regions that might similarly be distinguished. They recognised large discrete structural provinces, each one perhaps the remains of an orogen, a mobile belt long dead. Arthur Holmes was able to suggest the same kind of organisation was present in the Precambrian of Africa. Some of the earliest radiometric dating and structural considerations assured him that the structural provinces or old fold belts there had distinct ages and had arisen in a recognisable order. In later years much of what Holmes suggested for African geology has been confirmed. What was done for these shields has now also very largely been accomplished for the other great Precambrian regions of the world.

Not only have the Precambrian outcrop areas been assailed by the geochronologists and structural geologists but deep drilling through the cover of later rocks elsewhere in the continents has provided samples of Precambrian basement. The world-wide picture begins to emerge.

You will recall that the radiometric clock relies upon age-determinations of minerals taken, for the most part, from deep-seated igneous rocks such as granite. The age given is that of the final crystallisation of the rock. Once the crystal lattices of the minerals in the rock had set and the radioactive elements were locked in place slow decay began and the 'clock' was set ticking. In practice this means that the geochronologist can establish the time of the final cooling or uplift of an igneous body or a metamorphic belt. He will have a more difficult job finding out the dates of events that preceded this final falling of temperature. So what we record for much of Precambrian chronology is the age of the plutonic rocks rather than the dating of sedimentary materials or purely mechanical events.

When all the radiometric dates obtainable are collected together they bunch into rather well separated groups or clusters. This must

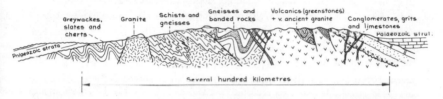

Figure 6-1 The Precambrian shields have complex anatomies, made up of many different kinds of rock and evolved through long phases of rock deposition and deformation. This diagrammatic section shows some of the kinds of relationships between the different formations but the vertical exaggeration used is very great.

mean that in a world-wide survey the plutonic events they represent were more common at certain times than at others. In their turn, the intrusions of big quantities of igneous rock mark orogenic phases when the mobile belts were being regionally metamorphosed into hard and unyielding and relatively immobile wedges. It is said that these periods or phases lasted about 800 million years. Obviously they mark important stages in the evolution of the earth, and they might be thought of as times when the chimes of the geological clock struck the hour. In this case the 'hour' is rather large, and it has been appropriately called a *megachron*. Megachrons are divided from one another by these great phases of widespread orogeny. Rather fortunately the Phanerozoic eon fits into this scheme so that we do not have to imagine that there is anything fundamentally different between Cryptozoic and later times.

Not unnaturally, groups of geologists working on the shields of different continents have erected their own calendars or tables of Precambrian events, incorporating local names for the various items or events that they can recognise. In each case what they recognised were cycles of events, with the formation of metamorphosed stable or immobile wedges and bathyliths occurring after long periods of geosynclinal or other deposition and succeeded by long periods of slow uplift and erosion. Many instances are known where the old sediment-filled troughs were folded and metamorphosed and invaded by granite not once but several times. When it is all over these parts of the crust seem to have become frozen and immobile, 'stabilised' we have called it.

Four such cycles have been suggested for Precambrian time reaching back 3,600–3,900 million years. The boundaries between them are drawn at the times when plutonic activity seemed to be at a peak. There are other suggestions as to where those boundaries should be fixed but the picture itself is not altered. What generally happened before 3,900 million years ago cannot be reconstructed from the evidence in the shields. Virtually all the earlier crust seems to have disappeared—or to have been reworked by the rock cycle in later times. At about $3,300 \pm 200$ million years ago at least five of the shield regions were intruded by large masses of basaltic rocks called anorthosites. The moon possesses rocks identical to some of these. Between this time and 2,500 million years ago basalt floods inundated what have been called the greenstone basins. Between 1,700 and 1,200 million years ago there was another bout of anorthosite intrusions from below into most of the shields. Anorthosite has virtually never penetrated the crust since then and what the conditions were that prompted it to arise where and when it did in those distant times remain unknown.

In order to see something of how, or rather, where, the Precambrian orogenic cycles occurred, we should turn now to a closer look at the shield regions of the world. That will give us more evidence to make a few further generalisations about early earth history.

Continental Cores

Not all the shields are known in anything like the same detail, but the realisation that they contain valuable mineral deposits has brought an increasing amount of attention to them during the present century. The best-known is surely the Canadian Shield, lying near to the great American mineral-hungry industrial centres. The Baltic Shield is

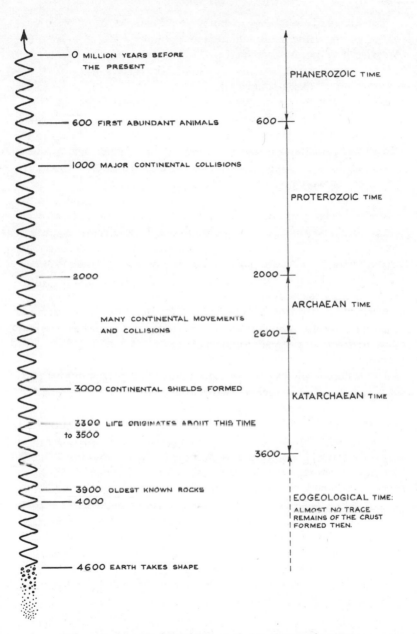

Figure 6–2   A table of Precambrian time and events.

perhaps similarly well-known, but is much smaller in outcrop area. Although the Guiana, Brazilian and Patagonian Shields in South America have not been studied to a comparable degree, it is certain that they contain vast reserves of metallic minerals.

Peninsular India is comprised of metamorphic and igneous shield rocks with later formations thinly spread on its surface here and there and gathered into local basins at its margins. While hilly eastern Australia is largely the remains of a Palaeozoic mobile belt, central and western Australia is a monotonous and arid landscape of shield rocks. There is no doubt that the Australian Shield is particularly well-endowed with metallic and other useful minerals.

The surface of the great African continent is composed almost entirely of shield rocks. Only in the Atlas Mountains of the north-west and in the Cape Mountains of the south are thick later rocks present. Elsewhere, near the east and west coasts for the most part, the crystalline formations are patchily covered by thin later sedimentary formations. Closely related and similar to the African Shield, one would guess, is the Arabian Shield.

Deep in the interior of Siberia, the Angara Shield has received much attention from Soviet scientists in recent years. It has no great continuous outcrop areas comparable to the African or Canadian shields.

The Antarctic snows and ice conceal most of the shield of that continent but Precambrian rocks have been found in many coastal areas.

In each of these regions several important facts are immediately obvious. First, the Precambrian rocks have all been subjected to prolonged uplift and erosion. Throughout much of Phanerozoic time they have probably contributed sediments to the shallow seas or deep trenches developing around them. The volumes of material that have been carried away during all this time are truly prodigious and one is prompted to wonder how it is that the continental uplift has managed to keep going. Second, all the shields are predominantly composed of metamorphic and granitic rocks: basaltic formations are restricted to a relatively few remarkable instances. Then there are the well-marked structural provinces and the widespread break between Precambrian and later rocks. Since Precambrian time the shields have remained unfolded and only little disturbed.

The Canadian Shield

The Canadian Shield offers the world's largest expanse of Precambrian rocks, occupying the greater part of Canada and virtually all of Greenland; it extends from Baffin Island and the Arctic Archipelago southwards on each side of Hudson's Bay and reaches into Minnesota in the U.S.A. Canadian Shield country is mostly a wilderness of conifer forest, muskeg and tundra, frozen for much of the year. Its treasury includes high grade iron, copper and many other ores, including the largest known mass of nickel ore. In the main the rocks are gneisses and granite. They include some of the most ancient known.

Precambrian rocks also occur in the great fold-belts at the edges of the American continent, the Appalachians and the Rockies, and in the Innuitian fold-belt high in the Arctic. In the mid-west and south deep boreholes in the cover of later rocks have reached the Precambrian basement and we have a good idea of the shape of its upper surface.

During the long history of the Canadian Shield there have been several periods when the compression and stabilisation of mobile belts took place, with all the accompanying metamorphism and intrusion that that involves. Although some parts of the shield, especially the south and east, are known in detail, much of it remains obscure. The old division of the Precambrian rocks here into Proterozoic and Archaean is still broadly usable.

The *Archaean* consists mostly of granite and gneisses, but there are local broad patches of sedimentary and volcanic rocks. There are some truly remarkable pebble and boulder beds in the Archaean, occurring in wedge-shaped masses up to 300 metres thick. They were probably deposited by torrents in large depressions in a hilly landscape.

Most of the sedimentary rocks—and the metamorphic types derived from them—are greywackes, the typical geosynclinal muddy sandstones. (More than 7,000 metres of these rocks have been found in some areas.) There are banded sedimentary iron ores, the silica-rich sedimentary rocks known as cherts, and slates with graphite. The chemistry of these rocks suggests that they were marine deposits and the graphite could be the metamorphosed remnants of matter that was once living, perhaps algae or other simple plants.

The metamorphosed rocks known as greenstones were once basaltic lavas and ashes, as much as 10,000 metres thick. Here and there they still retain the pillow-like masses produced during submarine eruptions. In view of these, we can assume that in several places there was volcanic activity on a grand scale; so grand a scale in fact that it suggests that something like an active oceanic ridge was not far away. On the other hand, the greenstone volcanics seem to have been deposited in definite sags or basins in the granitic crust. The earliest greenstone belts (about 3,400 million years old) were small, but later belts were several hundred kilometres wide and were filled with eruptives only in the course of rather long periods of geological time.

Two kinds of rock conspicuous by their absence from most of the shield are quartzite and limestone. The virtual absence of limestone may be due to the dearth of living organisms at that far-off time and to the presence of conditions which would prevent calcium carbonate being precipitated from water. This goes for other kinds of relatively water-soluble minerals too—there is no gypsum, salt or similar precipitates. Quartz-rich sandstones, or quartzites, result from sands that form where the less resistant minerals cannot survive the weathering and turbulence and where deposition is slow. The absence of Archaean quartzites supports the idea prompted by the abundance of greywackes, namely that erosion, deposition and subsidence were all rapid in those times.

Granite was intruded into these rocks on at least two separate occasions, and this is in keeping with the movements and vigorous changes taking place in the crust during Archaean times.

Proterozoic rocks generally seem to include less metamorphosed formations, but there are quartzites, slates, greywackes and other sandstones, conglomerates, limestones and dolomites. Volcanics are not so common as in the older division. Thicknesses of 20,000 metres have been measured in the Proterozoic rocks and several formations are separated by really large unconformities. In some regions, too, there are crowds of small igneous intrusions.

At the end of Archaean times the great Kenoran orogeny (2,500–2,600 million years ago) spread its convulsive effects throughout the shield, before the Proterozoic sediments were deposited. Within the Proterozoic eon the Hudsonian (1,700 million years ago) and the Elsonian (1,350 million years ago) orogenies took place and to cap it all came the Grenvillian orogeny (955 million years ago).

Using their analyses of geological structures and the thousands of radiometric dates available, Canadian geologists have been able to recognise several distinct provinces within the shield. Each has structures trending or aligned in a characteristic way and was formed within a fairly closely defined period of time. Some of these provinces are much better known than others. In the north-west the Bear and Slave provinces are rather small but may extend under much of the later rocks to north and south. The Churchill and Superior provinces are quite gigantic, enclosing Hudson Bay. The Nain is a relatively small region on the coast of Labrador while the Grenville runs from southern Labrador to the Great Lakes and New York. Beneath the sedimentary rocks of the Mississippi Valley a further structural province, the Mazatzal, has been recognised.

From such studies we can draw notions of how the Canadian shield may have formed. Possibly there were four primary continental nuclei to start with—slabs or 'bergs' of granitic material which had somehow arisen from the mantle. They have been called the Slave, Hudson, Ungava and Superior proto-continents and they already existed at the earliest Archaean time. We have no idea of their topography or geography, only a rough picture of their approximate sizes, outlines and positions. These proto-continents seem to have grown and merged to form a large stable mass by the accretion of new rocks in mobile belts at their edges, together with the addition of material, basalt, from the mantle below. Vigorous volcanic activity and orogeny must have made this a turbulent part of the earth's surface. Sediments were produced very rapidly by the decay and weathering of new volcanic and other rocks. During Proterozoic time comparatively long linear geosynclinal areas and broad depressions developed around the proto-continents and in regions now at the edges of the shield. At this stage much of the deposition of sediments took place at a slower pace than previously. Over thousands of square kilometres of shallow sea floor ironstones were precipitated by bacterial activity and not far away were barren rolling lowlands. Great out-pourings of basalt lava took place near Lake Superior, and in the north-west and in Labrador. During the Proterozoic upheavals granitic rocks crept or were forced up into the newer formations from below, but as the Cryptozoic eon came to an end the shield seems at last to have been free from these disturbances. It was no longer affected by intrusions of volcanic rocks or upheavals of a violent and startling kind but settled into a long, still-continuing, phase of gentle and intermittent rising as erosion gradually stripped away more and more of its surface rocks.

The Baltic Shield

Visitors from Scandinavia to the Canadian Shield are at once struck by the familiarity of the landscape. Underlying the superficial similarities of topography, vegetation, and climate is a basic geological similarity between the Canadian Shield and the wide area of Sweden and Finland in which lie the Precambrian rocks. The mountains of western Sweden and Norway are rather similar in part, for they too include metamorphic and igneous units in plenty. But the mountain backbone of Norway and Sweden is a Caledonian range, younger by far than the Baltic Shield to which it is attached.

Not surprisingly, the energetic Finns have most assiduously studied their Precambrian terrain. It has largely been scraped clean of overlying soils and sediments by the ice sheets of the Pleistocene period and the geology is displayed in almost embarrassing detail. The Scandinavian geologists, too, have recognised four great cycles or chapters in the ancient history of their land. Each one involved the accumulation of large thicknesses of sedimentary and volcanic rocks which were afterwards metamorphosed, folded perhaps several times and intruded by granite before the next chapter repeated the pattern.

Three structural zones have been found in the Baltic Shield, each containing highly metamorphosed rocks and granites. The Saamo-Karelian in the north-east seems to have evolved between 3,000 and 1,900 million years ago. The Sveco-Fennian in the south took shape between 2,300 and 1,500 million years ago and the Sveco-Norwegian belt in south-west Sweden and southern Norway came into existence over the period 1,200 to 900 million years ago.

In addition to these highly complex ancient rocks, the shield also reveals a thick and varied mass of flat-lying sedimentary strata, which are dated as between 1,800 and 1,200 million years old on the basis of the ages of intrusive igneous rocks. Somehow they have escaped the mangling and alteration that affected the other Precambrian formations.

A thousand kilometres away in the Ukraine an outcrop of Precambrian rocks emerges from beneath the cover of later sediments that stretch from the Baltic states across the steppes to the Black Sea. Sometimes it is referred to as the Ukrainian Shield but it is not a compact area of outcropping Precambrian rocks as are the other shields. Rocks from the Ukraine are said to give radiometric dates of 3,500 million years, and so must be as old as the earliest in Africa or the Canadian Shield.

The Siberian (Angara) Shield

With the exception of the Antarctic Shield, this is the least known (at least in the West) of the great Precambrian outcrop areas. Until comparatively recently it was relatively inaccessible and it occupies part of the continent that has not the happiest of reputations.

Pl. 9.   A pinch of the mantle may be indicated by these ultrabasic rocks of the Troodos
Mountains in Cyprus. It has been suggested that at the collision of the African and
European plates a small piece of the upper mantle may have been squeezed up between
the continents. The deep erosion of the surface since then has now revealed material that
is normally many kilometres down.   *Courtesy of P. L. Hancock*

Pl. 10.   Perhaps the most widespread of all eruptive rocks are submarine basalt lavas. The characteristic 'pillow' structure of lava erupted under water is seen in these ancient rocks at Newborough, Anglesey, in North Wales. These are of Precambrian age but virtually identical rocks may be found in all later rock systems. Some may have been formed at the edges of continents or near island areas, but others have formed in the deep ocean.   *Crown Copyright Reserved. Courtesy of Institute of Geological Sciences*

Pl. 11.   The *Glomar Challenger* has been used as floating drilling rig to obtain rock samples drilled from the ocean floor, and despite all manner of difficulties it has been remarkably successful. Our knowledge of the nature of the rocks beneath the deep waters of the world owes much to the skilful use of the many different technologies incorporated in this vessel.   *Courtesy of Global Marine Europa Ltd.*

Pl. 12.   Even the most familiar of mountains and glaciers, in this case the Rhone Glacier amid the Swiss Alps, still have much to reveal about the processes of mountain growth and decay, climatic change in recent geological times and about the delicate balance that exists between them.   *Courtesy of Aerofilms Ltd.*

Pl. 13.   The coastal bathylith of the Peruvian Andes. This granite complex is intruded into a vast accumulation of Mesozoic andesites and is not only the basis of the spectacular scenery but also contributes largely to the great mineral wealth of the region. It is also one of the largest bathyliths the world has produced.   *Courtesy of W. S. Pitcher*

Pl. 14. Asmara Mountain in Iran is an enormous elongate dome or anticlinal structure of limestones and other rocks. It lies in a region that has suffered great compression as the movement of lithospheric plates has gone on in Cainozoic times. Very many such folds lie within the area marked Ir in figure 1–5, and they include rocks that were once sediments on the Tethys Ocean floor.   *Courtesy of Aerofilms Ltd.*

Pl. 15.   Life on a coral reef 380 million years ago. From evidence of the fossils of the Devonian reef formations in various parts of North America this reconstruction shows something of the great profusion and variety of creatures living there. There are sponges, corals, lampshells, snails, trilobites, sea-lilies and octopus-like cephalopods together with fronds of seaweed and moss-animals. Unseen would be a host of planktonic and microscopic organisms about which we can only speculate.   *Courtesy of the American Museum of Natural History*

Pl. 16.   On the east coast of Devon Island in Arctic Canada flat-lying Cambrian rocks overly the grooved and rutted terrain of the very much older Precambrian. Few earth movements have affected the region since then and the rocks are almost continuously exposed in this cold desert area. To the right the sea is largely covered with pack ice.   *Courtesy of Geological Survey of Canada*

However, under the drive of the Soviet authorities, geological investigation of Siberia has had a high priority and has made impressive progress. This region of the eastern U.S.S.R. stretches from the Yenisei River in the west to the Lena River in the east and from the Arctic Ocean on its northern side to the latitude of Lake Baikal in the south. Like the Canadian Shield, it has conifer forest, muskeg and tundra, permafrost, extreme winter cold and other characteristics that did not endear it to early travellers.

In point of fact the Siberian Shield is revealed as a series of outcrop areas rather than one large region of Precambrian terrain. From time to time it has been extensively covered by thin Palaeozoic and Mesozoic strata, shallow water deposits for the most part.

Granites, gneisses and schists present in the shield are evidence of a long and complicated evolutionary history before sediments accumulated in the late Precambrian. In this, of course, it resembles the Canadian and other shields. This late phase dates from about 1,600 million years back and saw the spread of both sandy strata and limestones. Many of these beds contain great numbers of stromatolites, the limestone structures produced by lime-secreting algae. These large mound-like growths, a metre or more high, grew in the intertidal zones of the coast with the warm waves splashing between them, much as they do in Western Australia today.

By late Cryptozoic time the Angara Shield had passed through all the violent upheavals and orogenies that wrought so much metamorphism. It had been deeply eroded and planed off prior to the deposition of the thick blankets of sedimentary material. Then in the north-east, south and west it was again uplifted for a while before the sea encroached again and a new spread of shallow marine sediments slowly took place. By the end of this marine transgression much of the shield was once more covered with thin sedimentary rocks and time had advanced well into the Palaeozoic era.

Peninsular India
North of the great valleys of the Indus and the Ganges rise the Himalayan ranges, young mountain chains of enormous relief. To the south of the rivers roll the red plains and hills of peninsular India, a seemingly endless if rather subdued topography. To the traveller it is an unchanging, vast and ageless kind of countryside, somehow matched by the great geological stability of the region since Precambrian times. Here and there in the eastern and western coastlands younger rocks conceal the shield, but they are terrestrial or volcanic

deposits and have involved no large changes in the level of the land. Among these, the famous Gondwana formations, are sandstones, shales, and other rocks including coals, accumulated at intervals between late Permian and early Cretaceous times. South-west of Delhi and covering much of the north-western half of the shield are thousands of square kilometres of flat-lying floods of late Cretaceous and early Cainozoic basalt, the Deccan traps.

In this shield five major structural trend belts have been recognised. Most of their boundaries have not been mapped or studied in detail nor have many isotopic dates been calculated for the Indian Shield. Nevertheless, there seems to be a pattern to this shield that is in keeping with what is seen in others.

The oldest structural belt, the Dharwar, contains rocks 2,400 or more million years old and it lies in the south-west of the country. A wide variety of rock types there includes granites and gneisses and also volcanic and sedimentary rocks. A long and complex history to this parcel of rock formations was complete well before Phanerozoic times and very probably the Dharwar belt acted as a kind of stable nucleus to which other belts became attached. The Eastern Ghats belt is a rather similar metamorphic and igneous complex but the few radio isotope dates obtained so far hint that the belt may be not more than about 1,600 million years old.

From beneath the edges of the Deccan traps comes evidence of another belt, the Satpura. Analysis of radio isotopes from this belt indicates that it may have acquired its identity about 1,000 million years ago. North of this belt lies the Arawalli belt and imposed upon at least part of this is the Delhi, with a final orogenic episode about 750 million years ago. Between them these last two appear to complete the record of Cryptozoic orogenies and upheavals, but there remains yet another group of rocks which is rather puzzling. It is a further example of rocks that are difficult to date. Being only little deformed and resting unconformably on top of the Dharwar rocks, they might be as old as the rocks of the Eastern Ghats belt, or they might be much younger.

Finally, just south of the River Ganges is a group that offers the same puzzle as the so-called Proterozoic rocks in other continents. The Vindhyan strata are flat-lying shaly rocks with limestones, and with conglomeratic beds looking like river-laid deposits. There are even some small queer structures that might be primitive fossils. To the Geological Survey of India these rocks seem to be late Precambrian in age, formed largely of sediment from the metamorphic lands to the south. Without recognisable fossils, nor isotope dates, however, the age of this group is a mystery.

Africa

Precambrian rocks occupy more than half the face of Africa and elsewhere they are probably not far below the cover of late Palaeozoic or Mesozoic beds. In many ways Africa is geologically a unique continent and as far as the basement is concerned it appears on the one hand to be the most stable of all the continents while on the other to have suffered in its rift valleys the most remarkable fracturing. The mineral wealth of the African basement rocks has long been known, but only recently has exploration revealed many of the largest and most important resources. It is one of the few shield areas (South America and Angara are the others) where diamonds occur in their natural habitat.

Most African countries and the colonial regimes that preceded them have, or had, their geological survey organisations. Mining and trading corporations have also played a big part in geological reconnaissance. The result is that the outlines of African geology—at least south of the Sahara—were known by mid-twentieth century. Since then some areas have been studied in detail, and some have for one reason or another been quite neglected.

The ancient basement rocks of South Africa and Rhodesia in particular have been given as much attention as any. Some very important lessons have been learned from them and it would not be an exaggeration to say that they have influenced many of our ideas on early earth history. Rocks as old as 3,400 million years have been identified by isotope dating in Transvaal and Rhodesia. They include 17,000 metres of volcanic rocks seemingly floating in a sea of granite. The continental (sial) crust upon which the volcanics were erupted must have been extremely thin. So dense and 'basic' is much of the intruded igneous rock that we think it must have been emplaced at a very high temperature, high enough to indicate a nearly total melting of the upper mantle. It must have produced a fearsome upheaval and 'boiling' of the crust when it occurred.

Of all the many different rocks present in the African Precambrian, the most widespread and common seems to be granite. The oldest African granites are about 3,000 million years in age, and the younger granites are not essentially very different in nature or origin. When it comes to individual granite bodies there may be more than one mode of origin, but many if not most of the African granites appear to be bathylithic intrusions.

About 1,500 million years ago four regions had emerged as stable areas in the evolving crust of Africa. These were a large part of west Africa, two large regions in what is now central Africa, and an area now occupied by Rhodesia and the Transvaal. From this time on these regions have had virtually no severe geological disturbance. Their passage through episodes of metamorphic and granite intrusion was complete. Between 600 and 700 million years later widespread (the Kibaran) orogenies welded wide strips of metamorphosed granitic crust around the margins of the central and southern cratons. Then 400 to 500 million years after that came a further great sequence of orogenies known as the Pan-African. A hundred million years is an appreciable length of geological time, but even so, this orogenic phase brought a great transformation; it almost doubled the area of stable crust in Africa. The previously separate cratons and the newly heated and compressed mountain root regions between them were fused into a single shield. This was regional metamorphism on a truly grand scale. Apart from small areas in the north-west, south-east and the Cape region, the continent had achieved the outline we know today.

To summarise, the lengthy Cryptozoic history of Africa is divisible into early, mid and late phases. They correspond in a broad way to the general division of the Precambrian elsewhere. Early Precambrian events fell between 3,200 and 2,600 million years back, the mid Precambrian occupied perhaps a further 800 to 1,000 million years, and the late Precambrian extended from about 1,100 million years ago to the dawn of the Phanerozoic eon. The latest of all the basement rocks are sedimentary formations that may represent a story of unbroken deposition from the Precambrian well into Palaeozoic times, perhaps between 680 and 450 million years ago.

During each of these long eras the basement was taking on the structure it has today. The south central part of the continent seems to have acted as the nucleus to which large areas of metamorphic and igneous terrain were added in mid Precambrian times along the southern and western margins. Later addition involved the same kind of geosynclinal-orogenic cycle that is indicated by the structural belts in the other shields. At first sight this might seem to be a simple process by which the root zones of successive peripheral mountain belts are regularly added to the craton or core of the continent. It would be nice to record that it is so simple, but, alas, there are complications.

The South American Shields

Nature seems to do nothing by halves in South America. High mountains, dense forests and ancient rocks, all contribute to the extravagances of the South American scene. There are three shields where the basement lies at or near the surface in the eastern two-thirds of this continent. The northern shield occurs in the highlands of the interior of the Guianas and northern-most Brazil. To the south and west the Amazon basin intervenes, then the largest shield, the Brazilian, extends west from the eastern coastal highlands to the Parana basin. The Patagonian Shield is known only as a series of small outcrops emerging from beneath a cover of late Cretaceous sedimentary strata. All these regions are difficult to explore in detail and in consequence not much is known about the rocks.

The Guiana Shield, nevertheless, has yielded evidence of three long geosynclinal phases each followed by orogeny and the intrusion of granite. The oldest rocks are gneisses and schists. A mainly volcanic group with some sedimentary units, all very much metamorphosed and intruded by granites, makes up the Middle Precambrian here. Sedimentary and volcanic rocks belonging to yet another geosynclinal suite complete the list. One interpretation of this shield is that it seems to represent a succession of geosynclinal basins which began in the north. As time progressed, successive belts of the geosyncline were formed, each to the south of the previous one. So far, prospecting has not revealed many metalliferous deposits.

On the other hand, the Brazilian Shield has produced and still promises many bonanzas of minerals, metallic and otherwise. The oldest rocks are early Precambrian gneisses and granulites and on top of these is a very extensive body of geosynclinal rocks and volcanics. They are all much folded and contain many kinds of igneous intrusions. The later rocks of the shield are volcanic for the most part and also are severely folded.

The Australian Shield

Most of Central and Western Australia, an empty desert region for the most part, is given over to outcrops of Precambrian shield. A few flat-lying Phanerozoic strata cover it in places and there is a mantle of weathered rock, soil and sand that hides large regions of the basement surface, to the annoyance of prospectors. There is the usual array of desert land forms but the whole aspect of the land is of great stillness and little change from one eon to the next. This is indeed a reflection of the long staid history of the shield itself.

The Archaean includes granites, greenstones, gneisses and other metamorphic formations very like the Archaean in Canada. Australian geologists use the term in the same way as their North American colleagues. At least two great episodes of metamorphism and granite intrusion are known. They span the period 3,100 to 2,300 million years before the present with the radio isotope dates clustering around 3,100 and 2,650 million years. Most of the Archaean seems to have been involved in this later event, with enormous half-continent sized chunks of the crust being involved.

The Proterozoic has three very thick and widespread series of sedimentary rocks and the lowest of them rests on the worn stumps of the Archaean structures. Two further orogenic periods, 1,800 million years and 1,400 million years ago, separate the higher proterozoic series. The youngest Proterozoic series, the Adelaide, is overlain by Palaeozoic strata that are also part of this long phase of deposition. Most of the Adelaide rocks were deposited in a geosyncline and reach a thickness of about 15,000 metres. The series also has two other notable characteristics. It contains the fossils of strange invertebrate marine animals and has rocks that were probably formed under glacial conditions.

The Antarctic Shield
In its splendid and frigid isolation the Antarctic continent has managed to keep most of its geological secrets intact well into the twentieth century. The ice cap conceals all but a tiny portion of the continent, but during the last 40 years or so the geological outlines have become apparent. Using the sophisticated techniques of modern geophysical surveying, earth scientists have gathered information about the bedrock beneath much of the ice cover. Nevertheless, only in a narrow coastal strip can the geologists actually hit very much solid rock with their hammers.

In general it is the eastern half of the continent which has the largest showing of Precambrian rocks, but late Precambrian basement also occurs between the Weddell Sea and the Ross ice shelf. Igneous, metamorphic and sedimentary rocks are present and the eastern coastal area has yielded rocks thought to be 1,800 million years old. These formations have been intensely deformed, and testify to a long and complex history of regional metamorphism and igneous intrusion.

Then came a long period marked by the deposition of sediments in the eastern Antarctic. Early in the Palaeozoic era these too were disturbed by earth movements and somewhat metamorphosed. At the same time deep-seated igneous rocks were intruded into the region. Since then the Antarctic Precambrian basement has become warped and fractured on an impressive scale, so that in the mountains of eastern Antarctica it is raised to 3,000 metres above sea level. Few places in the world can match this record for an elevated basement.

Expanding Continents?

This quick survey of the shield areas gives a picture of successive mobile belts becoming welded on to proto-continental cores or nuclei beginning about 2,000 million years ago. No doubt it is over-simplified, but many geologists have found the idea acceptable enough and have thought of the continents as growing from an initial nucleus by the accretion of belts around the margins. As far as we can judge, mobile belts first began to appear only after the continental nuclei had become stable.

Each mobile belt began as a sedimentary basin or geosyncline. Then it became folded, compressed, and the deeper buried parts were metamorphosed and intruded by granite. The upper parts of these folded areas, the mountains and uplands, were removed by erosion until only the foundations and core were left. These crystalline regions were somehow now welded on to the old adjacent shield. The stable, stiff and inactive regions of the crust thus seem to have grown at the expense of the mobile restless regions. In turn a new geosyncline and orogeny would develop at the margin of this enlarged shield. Volcanoes would add material from time to time.

It is a nice hypothesis, neat and tidily explaining many of the features seen in the shields. Unfortunately some belts have been affected by overlapping orogenies and the isotopic ages found in the margins of the shield are not all younger than those in the centres. Later episodes of metamorphism tend to leave their characteristics superimposed on or masking completely the effects of earlier meta-morphism, and that may make for more difficulties for the geologist.

But have the continents in fact grown, stayed the same size, or diminished with the passage of time? During historical times vol-canoes seem to have erupted about 1 cubic kilometre of material per year. Some of it may be remelted crust but some must have come from the mantle. It is reckoned that at the present rate the entire volume of continental crust could have been erupted in a few thousand million years. Erosion of rock from the continents and the transport of the debris out into the oceans, however, do not reach this rate today, and the volume of sediments actually on the ocean floor at the present is relatively small. So, somehow the sediment eroded from the tops of continents is reincorporated within or under them. Here we have to envisage the rock cycle on a gigantic scale. How this cycle is brought about will be described in Chapter IX.

One group of geologists working in Africa has suggested that the later orogenic events in that continent involved materials that had already been involved in earlier 'earth-storms'. They believe that what we have described above as proto-continents or 'continental nuclei' are really remnants or 'islands' that escaped the later orogenies. And they have a different view of what happened in these orogenic episodes. The metamorphism and upheaval was not limited to old linear geosynclinal belts, but was a widespread upheaval and melting of the old granitic crust. If this kind of happening occurred, and the evidence for it is good, the Archaean African continent may have been at least as large as today's. Australian and Canadian geologists also find much in their own countries to agree with this. It leaves us with several new ideas to contend with, with doubts about the idea of continental accretion and for most geologists the uncomfortable feeling that we are still a very long way from understanding how the continents evolved far back in the Precambrian.

The Air: The Water

Among the many things learned about the world when the shields were still young is the fact that the geological cycle, the rock cycle, kept moving in much the same way as it does today. At least, many of its products, the rocks, seem to be quite similar. So no doubt, the atmosphere and the hydrosphere played parts in the cycle as important then as they are now. With signs of so many orogenies occurring in Precambrian time it is evident that the geological cycle was not moving at a slower pace than at present. If anything, the reverse might be indicated.

Many pages back the idea was introduced that both atmosphere and hydrosphere were produced from volcanic gases after the formation of the solid crust. Now it might be well to add a word or two about the evolution of those wet and windy envelopes of the earth.

The oldest water-laid sedimentary rocks are about 3,000 million years old. Being made of grains of minerals from existing rocks, they could not have originated without some kind of atmospheric weathering taking place. In these ancient sedimentary formations there are grains of pyrite, an iron sulphide mineral which soon oxidises when exposed to the air today. These ancient pyrite grains were broken from their source and transported to a different place before they were deposited with other grains of sand. The fact that they did not go rusty, that is, oxidise, en route rather indicates that there was little or no oxygen around at the time. Another gas or gases must have made up the bulk of the atmosphere.

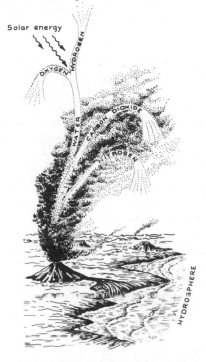

Figure 6-3 The original atmosphere of the earth has long been lost, replaced by gases released during volcanic activity. High in the atmosphere water vapour is broken down by solar radiation into oxygen and hydrogen. The oxygen remains but the hydrogen escapes. Many other processes which now affect the atmosphere cannot be shown on this figure. (After Spencer.)

Several planets have an atmosphere of ammonia, and possibly this gas was present in the early atmosphere of the earth. But it could not have been very abundant. Ammonia gas would make the hydrosphere rather alkaline and so favour the precipitation of carbonate rocks. No such abundance of carbonates—limestones or dolomites—occurs in the early Precambrian strata. These are among the good geochemical reasons for believing that the early atmosphere was relatively free of oxygen and other gases such as ammonia. The sources of most of the gases entering the atmosphere today are volcanoes. Most probably then, the young earth would have had an atmosphere rich in the gases given off by volcanoes and hot springs—carbon dioxide, carbon monoxide, nitrogen, sulphur dioxide, hydrogen chloride, and of course water. A carbon dioxide-rich atmosphere would, as we shall see later, probably be advantageous to the chain of chemical reactions that were needed to produce the first life on earth. The change to an oxygen-rich atmosphere came after the advent of life and the evolution of the processes of photosynthesis. We shall take a closer look at how this came about in our next chapter.

Once the crust of the primitive earth had cooled and the temperature of the atmosphere had dropped to below boiling point, rain would fall. Collecting in the great hollows and depressions of the surface it would soon have reached oceanic proportions during Precambrian times. It was no doubt close to its present volume shortly before the dawn of the Cambrian period. As it grew in volume, the ocean took into solution many soluble minerals and acquired the most characteristic marine feature. It became salty and, we noted earlier, its saltiness seems to have been relatively the same for at least the last 200 million years and probably much longer. Sooner or later most elements pass through the oceans in the course of their existence and a few remain there longer than others. Because they are so common and chemically so active, the elements sodium and chlorine dominate the soluble contents of the ocean waters. The chemistry of ocean waters today involves both inorganic processes and the activities of living things. More than one geologist has suggested that the long-pondered Precambrian absence of fossils, especially fossils with shells or skeletons, may, after all, have something to do with the chemical evolution of the sea. It is a matter that concerns us in the next section.

Precambrian Life

It is difficult to think of the earth as entirely desert, devoid of any form of life at all. Living matter is one of the most widely distributed yet extraordinary things to be encountered on our planet. Without living things earth would be very different and its history would have been appreciably different too. Where and when did life begin? Cambrian rocks have fossils which show that life had by that time already reached a high state of complexity. The origin and early evolution of life must begin back in the Precambrian eons.

In the history of our planet so far we have not referred at length to the influence of life and indeed it is only in the later Precambrian rocks that we encounter the first common signs of its presence. Widespread though life is today, it can exist only under a fairly narrow range of conditions and it needs the constant supply of a number of chemical elements and energy to keep it going. From the date of the oldest fossils to the present, life has been remarkable for the way in which it has kept going, has diversified in form and has, in short, evolved. The general feeling among geologists and biologists is that the earliest fossils stand not so much at the beginning of a long process of evolution as at the end of a longer and equally remarkable one.

All our studies of life tell us that it must have a long history indeed and that it probably originated here on earth at some remote geological date in the past. Although we do not know in detail how it took place, we have a number of hypotheses that can explain the stages necessary in the evolution of living matter from non-living chemical compounds. Of course there is plenty of room for argument and alternatives in this branch of science but there are many basic facts about which perhaps the majority of scientists concerned with the origin and early history of life are agreed. We should state a few of these at the outset.

Life as we know it seems to be confined to this planet in the solar system, largely because it can only exist within the particular range of conditions that are typical of earth. Mars and Venus may possibly possess little green men or other manifestations of life strange to us, as they have conditions different, but not too different, from those here. On other planets circling other suns in our galaxy there may be forms of life similar to, or more highly evolved than, our own. The possibilities never fail to arouse our interest.

But to return to life in its lowliest and simplest form on earth, we should recall that it is essentially dependent upon the chemistry of one element, carbon, which is in fact the twelfth most abundant terrestrial element. All forms of life today have many biochemical characteristics in common. To begin with they all contain numbers of giant molecules (macromolecules) of which carbon, oxygen and hydrogen are the essential elements. These macromolecules are also rather special in that they are polymers—built up of large numbers of simple basic sets of atoms. Two different kinds of macromolecules found in all living matter are proteins and nucleic acids. There are countless varieties of them grouped together into the basic stuff of life which we call protoplasm. Protoplasm occurs in discrete blobs or cells, and each cell is contained in a little bag or membrane. Before life could exist there had to be available the right materials and the right conditions for polymers and in particular the conditions for proteins and nucleic acids to form. Once they had been made, protoplasm and membranes could be produced. Given the right environment, cells grow and reproduce. They take in material from outside the membrane and incorporate it into their own structure. When a critical point is reached the cell may divide itself into two identical blobs, daughter cells each in its own membrane. This complicated activity is solely pursued by living matter. The non-living does not collect material from round about to build replicas of itself as the living cell does.

If we seek to account for the origin of life here on earth as the result of some kind of evolutionary process rather than as a single act of Divine Will, we have to ask whether these steps were possible at any stage during the early history of the earth. More and more evidence suggests that they were, and that the story is explicable in terms of the laws of chemistry and physics, together with another essential ingredient, time. Presumably life on other planets, or in other galaxies, would have resulted from a rather similar sequence of events. Carbon occurs in every planet in the solar system and in the known stars. Many meteorites reaching us from outer space contain carbon compounds.

Believing, then, that life could only have originated where there was carbon present, J. B. S. Haldane suggested over a quarter of a century ago that organic (i.e. carbon) compounds would slowly have accumulated in the early oceans. In time the concentration of organic compounds there would have reached a level at which all the reactions necessary to produce life could occur. He called these early oceans a form of hot, diluted broth or soup, and the idea has stuck with us.

The Russian scientist, A. I. Oparin, a few years earlier, had suggested that the right organic substances might be generated from carbon and other elements in the hot crust of the planet, perhaps in volcanic pools. These two theories called for something which seemed odd at the time but which now appears to have been a geological certainty. They required a non-oxidising, oxygen-poor atmosphere.

We know that several planets may have atmospheres poor in oxygen and containing methane, carbon dioxide, water vapour and ammonia. An experiment which seems to show that a whole range of organic compounds can be synthesised from such an atmosphere was carried out in the University of Chicago by a young chemist, S. L. Miller, in 1953. He passed an electric spark through a warm mixture of methane, ammonia and water vapour and found that these compounds combined to produce urea and several amino-acids. From non-organic material the simplest organic compounds were produced. Ultra-violet radiation directed on a similar mixture of gases and water vapour gave much the same results. In the early atmosphere on earth there would surely have been the gases Miller used and there would have been lightning and ultra-violet radiation from the sun enough to bring about the same kinds of chemical reactions. In time, perhaps, relatively large quantities of organic chemicals might be built up in the oceans. From then on there were suitable materials at hand to be moulded into compounds of increasing complexity until truly living substances took shape.

It was recently calculated by two geologists at Princeton University that this process (polymerisation) of building up complex organic compounds from methane and other gases could have produced an 'oil-slick' as much as 10 metres thick on the earth's surface. Others have thought that hot pools and volcanoes, which must have been common enough in Precambrian times, gave the right surroundings for the generation of amino-acids from simpler compounds. Some geochemists have suggested that shore lines and deltas where certain clay minerals are present to act as catalysts in organic reactions were essential. Wherever they were formed, such 'oil-slicks' would have been important sites for carbon-bearing molecules to evolve into more and more complex structures. Where they floated on the sea or coated sand on a beach these slicks would possibly have had access to salts which may have catalysed some of the reactions. They would also protect the water below from destructive ultra-violet radiation.

In time organic compounds in these environments were generated which had the advanced characteristic of collecting molecules from their surroundings and adding them to their own structure so that the basic pattern was reproduced over and over again. Even quite complex molecules would be captured, broken down into their component parts and these parts added to the structure of their capturing compound. One chemical 'ate' another. If the process continued long enough only a few 'successful' compounds would survive. When they were no longer collecting material from round about these substances may have themselves begun to break up into simpler compounds and so lost their self-reproducing abilities. There were possibly several kinds of 'successful' compounds resulting from different chemical evolutions. Life in fact may have had several different kinds of origin.

A milestone of major significance was reached when photo-synthesis first took place. In this process in plants the sun's energy is used to convert carbon dioxide and water into carbohydrates and oxygen. The green pigment, chlorophyll, is the essential chemical needed to bring the process about. How chlorophyll first originated is not known, but theoretically it could have been formed in several environments early in earth history. Incorporated into the structure of an early organism it would enable that organism to exist on a 'diet' of very simple materials and sunlight. Oxygen is a by-product of this activity, and to this day the green plants have retained the metabolic process of photosynthesis.

All this is very well, but it seems to be difficult to substantiate it by any sort of fossil evidence. Nevertheless, the Precambrian record has quite a lot to tell us about the early evolution of life. From limestones in the shields of North America, Russia, Australia and Africa fossils known as stromatolites have been obtained. They are rather cabbage-shaped laminated bodies of limestone or silica. Today such structures are produced by blue-green algae living in tropical tidal waters. The oldest stromatolites, however, are at least 2,700 million years old, and they must indicate that photosynthesis had been developed by then.

Other layered structures in Precambrian rocks are thought to be organic in origin but many objects of this kind once regarded as fossil are now known to be inorganic in origin.

In late Precambrian rocks in several continents fossil trails, tracks and burrows (they are known as 'trace fossils') suggest that soft-bodied creatures akin to worms inhabited the muds and sands of those days.

Actual fossils representing other marine creatures are very rare in Precambrian strata. Most of them are problematic or controversial, though there does seem to be a genuine case to be made out for worms or jellyfish-like organisms in Britain, Africa, North America and the U.S.S.R. In South Australia a quite remarkable collection of creatures is indicated by fossils from Ediacara. There are worms, possible sea urchins, jellyfish and other animals living on or in that ancient sea shore.

Perhaps the most remarkable as well as ancient fossils from the Precambrian are the splendidly preserved remains of algae, bacteria, fungi and other plants in shales and cherts in East Africa, Canada, Australia and Europe. The oldest of them, from the Fig Tree Series of Rhodesia, is about 3,400 million years old, and there seems little reason to doubt that photosynthesis was established by that time.

Amino-acids produced by living organisms have been chemically detected from rocks of this great age and in other parts of the world similar 'chemical fossils' have been discovered.

All these fossils and traces of very ancient life are well and good in their way, but compared to the fossils of the Cambrian they are rather

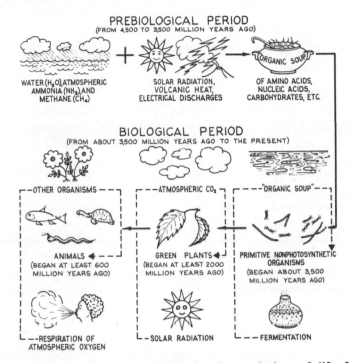

Figure 6–4 Some possible steps in the evolution of life from non-living matter. The sources of energy are shown at the bottom of the diagrams and at centre top. (After A. L. McAlester.)

primitive. Except for the stromatolite-making algae, not one of the authentic Precambrian fossils has a shell or skeleton of calcium carbonate. In the early Cambrian rocks we have over 450 species of fossils, the majority of them sporting some sort of hard shell or skeleton. Such a sudden burst of armour-bearing marine animals represented something new in ancient biology.

Several ideas about this have been tossed around over the years. Some geologists suggest that the chemical composition of the seas and atmosphere was different to the extent that until Cambrian times the sea was too acid to allow the formation or preservation of carbonate shells. Others maintain that organisms themselves were not capable of carrying out the necessary chemistry to secrete calcium carbonate or phosphate for hard supporting or protecting parts. No doubt the fact that the atmosphere was oxygen-poor in early Precambrian times but progressively increased its oxygen content once chlorophyll-bearing plants arrived (about 3,000 million years ago) has much to do with this Cambrian feature. Oxygen had risen to occupy 1 per cent or more of its present level in the atmosphere by the dawn of the Cambrian period.

Amongst the economically valuable rocks of the shields are the banded ironstones, mined on such a large scale in North America and Australia. These rocks consist of alternations of iron oxide, clay minerals and chert or silica. In the general absence of atmospheric oxygen at the time these sediments were deposited, their formation is perhaps best explained on biological grounds. Organic activity in broad shallow lagoons or lakes may locally have given rise to a chemical environment in which iron and silica were precipitated. Algae would be likely candidates to produce the locally oxygen-rich waters and atmosphere necessary for the iron minerals to be precipitated from the waters. Certain bacteria do this in modern lakes and swamps. The banding may represent periods of precipitation followed by intervals necessary for the chemical concentration to be re-established.

Animal life is vigorous and active compared with that of plants. To sustain its activity animal life produces energy by oxidising its food, so it needs a good supply of oxygen. Perhaps not long before the Cambrian period began the plants had raised the amount of free oxygen sufficiently to allow animals to use it in their metabolism. Up until this time perhaps all animal life had been limited to unicellular creatures. Multicellular animal life was a consequence of there being enough free oxygen available to sustain rapid metabolism and it may first have flourished in the vicinity of the oxygen-producing marine plants. Once metazoan animal life became possible the exploration of other bio-chemical avenues such as shell formation, was soon taken up.

Until this late Precambrian date the earth's surface would have been dosed daily by strong ultra-violet radiation. The ultra-violet waves are destructive to life and it has been suggested that life could only have persisted in the lower parts of the sunlit zone of the sea while this radiation was in force. Today's oxygen-rich atmosphere has enough ozone in it to filter out the harmful ultra-violet radiation. Ozone, an inert gas with molecules consisting of three oxygen atoms, is produced in tiny quantities during photosynthesis. As the biologically produced free oxygen level rose so did the ozone level and the UV radiation effects were correspondingly diminished. More ocean waters became available for occupation.

So the new levels of free atmospheric oxygen set the stage for the emergence of new organisms using the more rapid and effective oxidation processes of metabolism. Organic evolution now utilising photosynthesis on the one hand, or oxygen-based metabolism on the other, was at the end of the Precambrian poised to take advantage of the many environments that were thus made available at the beginning of the Palaeozoic era.

There was a further possible side effect of this depletion of carbon dioxide from the atmosphere by the activity of photosynthetic plants. The atmosphere no longer retained the infra-red radiation received from the sun and in consequence lost heat. With the removal of the 'greenhouse' effect as this heat-capturing mechanism is called, the climate may have become colder, more extreme. An ultimate consequence of such a train of events could be a glacial period with the growth of continental ice caps.

Freeze-up

From some of the latest of the Precambrian sedimentary rocks to be deposited on or around the shields in North America, Europe, Asia, Africa and Australia comes evidence of a Precambrian ice age. The rocks in question possess all the features we associate with present-day moraines, glacier muds, outwash gravels and the banded deposits of glacial lakes. The ancient conglomerate beds contain faceted, polished and striated pebbles and boulders, exactly comparable to modern instances. Moreover, these highly variable rocks rest locally on polished and scratched surfaces which reflect all the features of the recently ice-covered and eroded rock pavements known so well in many parts of the world. A late Precambrian, sometimes called Eocambrian, ice age must have occurred, involving a large part of the earth's surface. To judge from the thicknesses of glacial deposits in these rocks it was an ice age at least comparable in duration and severity with the one through which the earth has been passing in the last million years or so. Probably it was much longer, a succession of ice ages, not exactly contemporaneous on every continent.

The picture is one of a world devoid of vegetation, much of it in the grip of snow and ice, its mountains shrouded and its seas locally under pack ice. The effects of this extreme climate must have been felt to some degree in every part of the world, even if it were only a slight lowering of temperature at the equator. But what this drop in temperatures must have meant to the living things of the day is hard to imagine. Little is known of the life of those times, except that there certainly were organisms of various kinds in the sea. Had they been accustomed to life in a warmer regime and were there mass extinctions because of the climatic change? So far the answers elude us, but with a remarkable suddenness, it seems, varied and vigorous groups of organisms were to appear in the seas of these regions shortly after the glaciers had melted from the scene. In years this interval may have numbered several million, but it involved one of the most profound changes of scene and significant steps in organic evolution imaginable. The animals that were now to appear possessed skeletons and more complex anatomies than previously. The sudden arrival of such fossils marks the beginning of the Cambrian period.

There may even have been an earlier ice age if we can so interpret a much older and less easily decipherable group of Precambrian rocks. But the evidence is not wholly convincing.

The cause, or causes, of even the most recent ice age is not fully understood, so to find an explanation for a Precambrian glacial age is even more fraught with difficulties. It may have been due to a diminution in the amount of solar energy reaching the surface of the earth, or to a change in the circulation of the atmosphere and oceans from a more local cause. A suggestion, noted above, that has found favour is that because of the widespread and long continued activity of the blue-green algae in the late Precambrian seas the proportion of carbon dioxide in the atmosphere dropped appreciably. The activity of the algae is shown by the immense numbers of late Precambrian stromatolites, the calcium carbonate mounds built up in the intertidal zone by these plants. Carbon dioxide was necessary for the precipitation of the carbonate and removing it from the air reduced the infra-red retaining capacity of atmosphere to the extent that the 'greenhouse effect' disappeared. Once a cover of snow or ice had spread upon the earth it further reflected the sun's energy to make matters worse, and an ice age was produced. Only after a long spell did the grip of snow and ice give way, and the reasons for that are no less easy to postulate than the causes of the original cooling. A redress of atmospheric composition would scarcely be likely, so other factors may be more important. The puzzle remains.

CHAPTER VII

On the Shelf

BETWEEN the ancient shields on the one hand and the relatively new mountain ranges on the other lie broad areas of the continents that have very distinct characteristics of their own. These are the regions in which the old Precambrian basement is veiled by a thin cover of comparatively undisturbed later sedimentary rocks. The great plain-lands of the world, the prairies, the pampas, the steppes, are obvious examples, but the great interior lowlands such as the Mississippi valley, the Amazon basin and others also belong to this kind of terrain. The interior of North America, for example, clearly has two obvious divisions. To the north lies the Canadian Shield, wooded rough country; around the margins of the shield and to the south a veneer of sedimentary rocks gives rise to open, gentle slopes and arable land, prairies and plains.

Geologists from Lyell onwards have commented that while the geological behaviour of the edge of a continent may be periodically very restless, the interiors seldom seem to do more than rise or fall gently, as if the continent were taking a slow deep breath. Although the shields were areas of geosynclines, mountain-building and igneous activity in the distant Precambrian eras, they have been conspicuously quiet and stable ever since. Only the broadest of uplifts, depressions and tilting movements have affected them. During the Phanerozoic eon they have alternately been covered by shallow seas and exposed to widespread weathering and erosion. The sediments left behind by the seas, lagoons and rivers have generally been thin, resting upon the eroded surface of the basement rocks or upon eroded remnants of previous sedimentary rocks, the legacy of previous seas.

In some parts of the world erosion has stripped away any Mesozoic and Cainozoic strata there may have been, and the Palaeozoic deposits are revealed. When these, too, are weathered away the record of the incursions of the sea across these regions of the continent will be lost forever. It is, in fact, now difficult, if not impossible, to say how far those seas were successful in flooding and covering many of the continental shields.

A short acquaintance with some of these rocks reveals that although they may look the same over wide areas they may change their thickness. In the space of a few miles a group of, say, limestones may become several times thicker. Elsewhere they may thin out and disappear between other rocks above and below. These lateral changes testify that some areas subsided more than others, and that locally there were islands or regions that stayed awash or even had a tendency to rise from the waves. Depending on their shape, the rising areas are known as domes, arches, axes or uplifts and the dip of the sedimentary rocks away from them has largely developed as a result of gentle vertical uplift over intervals of millions of years. Some of these structures, and the basins or troughs between them, cover thousands of square kilometres. In North America, for example, the Transcontinental Arch stretches from near the Great Lakes south-west to Arizona. The Congo Basin in Africa is virtually the size of the present Congo River drainage basin, perhaps a fifth of the entire continental area.

Where the basins and troughs were reached by the sea, large bodies of water penetrated into the interiors of the continents. One can imagine that these may have had a moderating effect upon the continental climates and offered attractive habitats to many kinds of animals and plants. When they dried up the biological consequences would have been just as important, though disastrous for many of the inmates.

Not all such basins were, however, connected to the sea. One of the largest and most important continental basins from the point of view of our story is the Karroo Basin in South Africa. It first took shape in late Carboniferous time and lasted about 60 million years, until the late Triassic period. It is filled with fluvial, lake and swamp deposits including coals. At the end of this time were great outpourings of basalt in the region, when lava flows covered much of the basin to a depth of 1,000 metres.

One important thing impresses us from the histories of these continental interiors. It is that although the basement of granitic and metamorphic rocks has warped and flexed gently from time to time the movements have been rather slow and the overall behaviour of a continent has been rather like a rigid slab of toffee. The old crystalline basement will bend on occasion, but only if the pressures are sufficient. Otherwise it stays inert, perhaps cracking a little here and there without changing its shape appreciably.

It struck the geologists of two or three decades ago that this kind of existence seems to be an odd thing if the continents had been drifting over or through a 'sea' of sima. How had the basement remained so inert and steady during its bodily translation from one place to another? It seems more easily explained now, but for some time it puzzled earth historians, and they had no generally accepted ideas to account for it.

At this stage it is worth digressing slightly to observe some of the geological processes at work while the sea is spreading across areas of continental size. Then we can take a case history, say the stable interior of North America, and see how it evolved in Palaeozoic times. A glance at the evolution of the Karroo Basin will also reveal some interesting facets to the history of continental interiors largely cut off from the sea.

Waves Upon the Shore: Marine processes

Nowhere is the continuing change to the face of the earth more impressive than along coastlines. Here the assault of the sea upon the land is brought home to us with every tide, and emphatically with every storm.

Many of the world's coastlines are relatively free from storms, strong currents or high tides but others suffer from the unremitting attack of gales, extreme tides and a rapacious undertow. Each year property is destroyed and lives lost as the sea batters away at cliffs and shores. In exceptional cases, as in the typhoons that inundate the land around the Bay of Bengal, the destruction wrought by the combination of high tides, storm surge and heavy rainfall is truly catastrophic. While the daily splash of the waves against the land keeps up an incessant nibbling away of rock, sand or soil, the rare storm may carry away the land in great bites. To the catastrophists of the eighteenth century such storms offered puny examples of the kinds of geological events that were possible in the past ages. No one will deny the tremendous geological work done by storms, but it is by the regular activity of the sea that the greatest effects have been achieved in the long run.

What it all amounts to is that the edges of the continents are being eroded by the sea, its waves and tides, its currents and other movements. The debris from this erosion is usually carried some way off from where it was formed and is deposited in sheets, dunes, banks and other variously shaped masses that may in time become sedimentary rock. It is worth taking a moment or two to observe some of the physics behind this most essential geological process.

We have all seen corks or boats bobbing on the water and although they may move up and down as a wave passes they seem to remain essentially in the same place. This surely means that the water itself is primarily moving up and down and not laterally, otherwise the floating objects would be swept along with the waves.

Closer observations show us that the water in a wave tends to move in a more or less circular path unless it meets an obstacle at depth. In shallow water this circular motion may be prevented and the wave begins to 'break'. Waves are generated by the effect of wind or some other disturbing factor upon the surface of the water and they move in the same direction as the wind. This is the direction in which the circular motion takes place and particles of water moving forward from the crest of a wave turn under and return to the surface at the crest of the next wave, unless there are currents involved. The strength of the movement diminishes downward and at a depth equal to about one-third to one-half of the distance between wave crests the water is virtually unaffected. You can see from this that below a depth of a few metres the ocean is normally a quiet place. This is why travel by submarine is so comfortable when the ships at the surface are being thrown around violently by storm waves. Of course there are rare occasions when large storm waves may stir up the sea floor at great depths. For the most part, however, the effect of surface waves on the sea bottom is important only in shallow water, generally near to land.

'Breakers' are another matter altogether. As we watch a wave approaching the shore we see that when it enters water with a depth of about a third to a half of the distance between the wave crests, it begins to 'drag' on the bottom. The crest moves on faster than the base and spills over as a tumbling mass; surf is exactly this, on a large scale. The tumbling water continues its forward movement and rushes up the beach or pounds against the cliffs. Here is the essential activity in marine erosion—the erosive power of breakers—the combination of the hydraulic force of the water itself and the effects of the rock particles picked up from the bottom and thrown against the land.

Now coastlines are by no means straight lines, or tidy places where the forces of erosion operate everywhere equally. Headlands tend to receive more battering than the bays they protect. Waves approaching the land from out at sea are relatively straight and in parallel lines. Hitting a headland or rock, or meeting a shallow patch, they bend and become deflected. The headlands receive the brunt of the attack. At the beach and on gently sloping shores the water rushes up onto the land until its energy is expended. Then it has to fall back, and the undertow and backwash are produced. Waves may hit the beach at an angle but the backwash is always directly down the steepest angle of the slope of the beach. Oblique uprush, or swash, and direct backwash have the effect of moving sand and sediment along the beach. Debris worn from the headlands of rugged coasts is swept into the quieter bays, perhaps in the form of spits and bars.

On many coastlines erosion is obviously the dominant process. These coastlines share a number of interesting erosional features, which may impart a degree of charm or even splendour to them. Because waves break at or just above sea level, their erosive action is limited to the zone just above and below the water surface. Where there is a big tidal range this level of course varies with the tides.

The first effect to be produced by erosion is for a notch to be scoured out of the shore at sea level. It may be cut back into the land until the overhanging mass collapses and is washed away. As the process continues a cliff is produced. In most, though by no means all, cases the cliff ends abruptly just below water level in a wave-cut platform gently sloping to seaward. As the cliffs recede the platform becomes wider, but this means that the waves have to cross an increasing expanse of shallow water to reach the cliffs. As waves cross the platform they lose their erosive power and the rate of cliff formation dwindles with time. From place to place the rate of erosion and the resistance of the rocks varies so the relief of the cliffs may change along the coast. There may be caves where local weaknesses have been probed in the foot of the cliff. Sea stacks may remain as erosional remnants as the cliff-line is pushed back. The variations are endless.

Erosional features such as cliffs and stacks formed in past geological ages are not commonly preserved, but they are found in a few places. Better known and relatively common are the sands and shingles of ancient shorelines, now incorporated in the sedimentary formations that drape the shields and continental margins.

Beaches occur on most coastlines, either as sandy patches or as longer accumulations where there is little active erosion of the bedrock. Where the energy of the waves and currents fails, rock particles, shell fragments and other debris are deposited as sand or mud. Unconsolidated loose masses that they are, beaches may be destroyed or formed rapidly by storms. Some are formed as submerged ridges or accumulations, called offshore bars, parallel to the beach itself. Sand may be gathered in these until it ´builds up above water when it forms a barrier beach. Where they extend from headlands or stacks strip-like accumulations of sand may form spits.

Beaches are the narrow exposed edges of the sheets of sediment that extend out to sea from most coastlines. At the shoreline where the expenditure of energy by waves and currents is very large, so are the sediment particles that have been broken and moved by the water. Further from the shore in quieter water the sediment is sand, silt or mud. In the stillest regions only a little fine sediment may be accumulating and some of that may be wind-borne dust.

Given enough time and stable conditions, i.e. little change of land or sea level, it seems that erosion and deposition tend to smooth out the near-shore sea floor until it is a relatively plane surface rising very gently towards the land. This slope is generally concave upward with the steepest part near the shore, where it may merge with the wave-cut platform. If the sea level rises there is increased space above the wave-cut platform for the action of waves and the cliff-line is pushed landward with the renewed vigour. Rising sea level may tend to drown the land, forming irregular rugged coasts. On the other hand, regular, very gently seaward sloping coasts may be formed by a recent retreat of the sea from the land or by the late accumulation of land-derived sediment.

Our excursion to the seaside has had an objective, namely to look at the process of marine erosion and to observe how the sea is constantly trying to extend its wave-cut platform. The material eroded from the land may temporarily form beaches but much of it may eventually be carried into deeper calmer waters to accumulate as sheets of sediment. With time, this material collects layer upon layer and is compressed by its own growing weight into sedimentary rock.

A rising sea level drowns the land and allows marine erosion to extend over ever-increasing areas. Behind the attack of the waves comes the deposition of sediment and as the beaches advance across the newly cut platform so the zones of deposition behind them similarly extend their area. This advance of the sea and the spread of successive blankets of sediment is called a *transgression,* and transgressions have been in progress on and off throughout most of geological time.

When the sea retreats because sea level is falling, or because of local earth movements raising the land, a *regression* occurs. Under these conditions sediment is stripped away or perhaps marine conditions are replaced by estuarine, deltaic or terrestrial ones. Naturally,

the sediments and the organisms within these environments differ
from those laid down in the sea.

Floods and Basements

The flooding of the continents by the oceans seems to have been a
ding-dong affair throughout geological history. At one time the
continents, especially the stable Precambrian Shield and the moun-

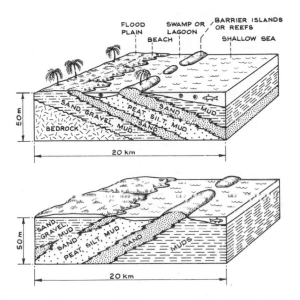

Figure 7-1   As the sea encroaches upon the land (*above*) its progress
is marked by the landward migration of sea floor and beach deposits.
The transgression is marked by migrating environments. In a regression (*below*) the migration is in the opposite direction. From the
earliest geological times there have always been these movements, no
continent has ever had static coastlines for long.

tains, were emergent, the sea in retreat. At others the oceans seem to
rise inexorably, flooding more and more of the margins of the land,
extending marine erosion over wider areas and depositing greater
volumes of sand or other sediment. One is prompted to wonder if
there ever have been occasions when the continents were dry right to
the edges of the continental shelves or whether the lands have ever
been completely submerged. What might have happened to influence

things one way or another? Some will argue that the volume of the sea has been increasing over the geological eons: other geologists believe that it has remained essentially the same. On balance, the latter seem to have the greater evidence in support of their case.

Nevertheless there do seem to have been times when transgressions were occurring over enormous areas. Indeed there are a few occasions in the Phanerozoic eon when transgressions were in progress on every continent. The Cambrian, Ordovician, Devonian and Cretaceous periods provide such examples.

In late Silurian, Permo-Triassic and early Cainozoic times the reverse seemed to apply and there were notable examples of regressions to be found across the world. If the actual volume of ocean waters did not change, there must be other factors that influence these happenings. In glacial periods the volume of the sea can actually be affected by the growth of ice on the land, but we can ignore that here. No great transfer of water to ice-caps occurred in Silurian, Permo-Triassic or early Cainozoic times.

Another possibility is that the continents have somehow been raised up, as if inflated from below by some great bathylithic event. That scarcely seems likely but is perhaps not impossible. Or perhaps the actual volume of the ocean basin itself, as distinct from the volume of water, has changed. Until recently it was hard to see how that could have happened, but the discovery of the volcanic oceanic ridges has given us some new ideas about this possibility. Later on we shall return to this point. Locally too, the events in geosynclines may have had effects upon the level of land and sea round about.

Earth movements of a less slow and widespread kind within the shields and over the sediment-covered extensions of the shields are known to occur. Isostatic uplift occurs in shield areas as well as in mountainous regions. After the great load of the glaciers was melted from the Canadian and Baltic Shields at the end of the Great Ice Age the ground slowly began to return to its pre-glacial level. It had been pushed down by scores of metres under the weight of ice upon it. This surely means that far from being absolutely rigid the basement is capable of plastic deformation and can return to its previous shape when a distorting influence is removed. The proposition that great weights of sediment could set the growth of geosynclines in motion in a similar way was considered by geologists and favoured by many of them a hundred years or more ago.

We remarked in an early paragraph in this chapter that the Precambrian continental floor, the basement upon which later formations have been deposited, has been warped into broad basins and arches, domes and troughs. Just how or why this happens is not certain, even where we have been able to follow the history of the arches or basins back to their inception. Many geologists have thought that there must be some cause lying beneath the sialic rafts of the continents that is responsible for the ripples on the surface of the sial. Convecting currents within the sub-continental mantle, according to some earth scientists, might have so eroded away patches of the lower sial that its upper surface sinks locally below the general level. Arches perhaps in the same train of events may result where sub-sialic movements tend locally to thicken the lower levels of the continent. However, it is not easy to see an appropriate pattern in the arches and basins of most continents in the light of this hypothesis. Other geologists have formed the idea of local, though giant, bathylithic intrusions at great depth influencing the contours of the shields. These bathyliths would be made active, hot and expanding by local disturbances in the mantle beneath. There are several variations on this theme.

Another proposal has been to credit the earth with enough loss of heat to cause substantial contraction during recent eons. As a result the continents contract by plastic deformation and also by brittle fracture, producing ups and downs in the outer surface as they diminish in size. Although we can find basins with Precambrian basement floors that are indeed fractured and which suggest compressional structures of a large size, they do not seem to provide much real basis for the idea of contracting continents.

North America has several well-known arches and domes south of the Canadian Shield, and a few to the north of it too. They seem to have originated mostly in the Palaeozoic era, with little movement taking place in the Mesozoic or Cainozoic. In Ordovician and Devonian times nearly all the continent was covered by transgressive seas and again in the Cretaceous this occurred. The relationship between transgressions and basement movement has been a longstanding puzzle to which only recently has a possible partial solution been offered.

Although every continent has its basins and domes, upwarps and downwarps in the basement, not all continents have experienced the same dowsings by the oceans. Much of North America, for example, was flooded during the Palaeozoic era and again in the Cretaceous. Not so in the case of Africa; most of it seems to have stood well above the waves throughout its history. Its biggest basin, the Karroo, is in fact filled with continental, largely terrestrial or swamp deposits. The reason for the remarkable stability that Africa has maintained is, at least, elusive.

The North American Interior

Between the Canadian Shield, the Cordilleran Geosyncline and the Appalachian Geosyncline the basement rises and sinks as a number of arches and domes that played a significant part in the Palaeozoic history of the continent.

In northern Alberta is the Peace River Arch; the Transcontinental Arch extends from Minnesota to Arizona and in Montana is the Montana Dome. The Ozark Mountains lie on the site of a dome and from Nashville, Tennessee, north to Michigan lies the Cincinnati Arch: Between Peace River, north-west Canada, and Montana and occupying much of Saskatchewan is the Williston Basin. Michigan lies four-square upon the Michigan Basin, while much of Illinois and Indiana is underlain by the Illinois Basin. Most of these broad, gentle features developed during Palaeozoic time and have been dormant ever since. They show the surface of the basement warped hundreds of metres from its general level and they seem to have been important features in the Palaeozoic geographies.

From what we can see of the Cambrian and early Ordovician rocks, these warps were not present in the basement then. At the very beginning of the Palaeozoic, the Precambrian basement was laid bare over the whole of the continental interior. A long period of weathering, perhaps as much as 50 million years, elapsed before the Cambrian seas invaded the continent and a lot of weathered rock and soil lay ready to be washed into the Cambrian seas. During early and mid-Cambrian times sea level rose steadily, or the basement sank uniformly, until the waters spread across the entire continent. It must have been a remarkable sheet of water for although it was so extensive it seems to have been nowhere as much as 30 metres deep. There were numerous islands and shoals of Precambrian rocks breasting the waves and tides. Everywhere near the exposed shield the sea floor was sandy with material washed from the old landscape, sifted and winnowed to and fro by waves and currents until only the purest quartz sands remained.

Farther south from the shield the water was perhaps a little deeper, a little less agitated. It was clear and warm and from it was precipitated calcium carbonate in great quantities. There are many fossils in the limestones that this carbonate produced. A large proportion of them have calcareous shells or skeletons, and limestone-secreting algae are also present. It is tempting to think that much of the carbonate was precipitated by organisms of one kind or another but so far this has not been proved. The limestone beds that have resulted from these deposits are nowhere very thick over the continental interior but near the geosynclines in east and west they reach 600–700 metres in thickness.

Careful study of these ancient limestones and their fossils shows that there were several different kinds of environment in those shallow seas. Those close to shore were sandy, scoured by currents and tides; those elsewhere deposited in less agitated waters were free of sand. We can trace how as the early Palaeozoic times passed the seas and the belts of different sea floor sediments advanced slowly northwards over the shield.

At the end of the early Ordovician epoch the gains won by the sea from the land were rapidly lost, uplift of the shield or the lowering of sea level affected thousands of square kilometres of the continent. Deep gullies were cut in the recently raised early Ordovician rocks before the sea returned. The landscape must have been barren, rocky, white and grey in the glare of the sun high overhead. As it advanced from south to north the land-hungry sea deposited once again a thin, persistent sheet of quartz sands. Farther out to sea the clear waters were the scene of great activity on the part of sea creatures and plants and the slow deposition of layers of carbonate mud.

By late Ordovician time the entire continental interior was submerged: from Mexico to the Arctic Islands and Greenland the sea spread out, shallow, clear and warm. In the west it merged with the deeper waters of the Cordilleran Geosyncline and in the Appalachian area earth movements had given rise to a land mass from which mud and sand were now coming. There were shallows, shoals and lagoonal regions, islands and reefs. Because the water was so shallow and the relief so low, slight changes in water level made big differences to the pattern of land and sea. The intertidal flats must have been enormous. The shoreline must have been constantly changing, while the currents and waves would have responded to changes in topography and wind direction. The skeletons of the myriads of small marine creatures were strewn over the sea floor and piled into layers and sheets that have become the fossil-fragment limestones that so delight the palaeontologists in North America. Despite the extent of the seas to north and south, the Transcontinental Arch seems to have existed as a chain or barrier of desert—or at least barren—islands.

Then came what we may regard as another dry spell. The early Silurian was a time when the sea withdrew once again to the north and south before returning in full flood to cover virtually the whole continent again. Throughout this active period the Transcontinental Arch generally separated the seas of the Mississippi Valley and the Williston Basin. The Ozark Dome and the Appalachián fold belt similarly stood out against the waves.

The Silurian seas here were warm, clear and relatively free from sand or mud. They offered an attractive habitat for many invertebrate animals amongst which flourished the corals, algae and other reef-building organisms. The reefs of the Middle Silurian of the Great Lakes–Michigan area, as in western Europe, were ridges or mounds rising quite steeply from the bottom to the surface of the water. Within the reef various kinds of small marine creatures built up a rigid limestone structure or framework. In the spaces enclosed by this framework lived many other shelly creatures such as brachiopods, snails, clams and sea lilies. The framework also acted as a natural trap for the remains of dead animals, shells and so on. At the surface the reef was planed off flat by the waves while its flanks were alive with creatures or draped with layers of debris worn from the top of the reef by the sea.

The reefs were several metres in height and in plan were rather irregular patches. To a considerable extent their shapes must have been influenced by the directions of currents, by the wind-driven surface waves, and by the way in which the sea floor subsided. When the middle Silurian reefs of the American mid-west are plotted on a map they outline the Michigan Basin and must have formed a great archipelago here at that time. Later on, in Devonian times, coral and algal reefs were to form continuous ramparts lining the coast, like the barrier reef off Australia today. Along the edges of the Williston Basin in Manitoba the scene was very similar.

In late Silurian times there was a shallowing of the seas across North America and they may have withdrawn completely from several regions. To the north-west and in the east large expanses of the sea were virtually cut off from the open water. Under the hot, arid climate these giant lagoon-like areas acted as great evaporating basins. The waters became too salty for life and in due course they began the precipitation of deposits of rock salt. In the Michigan Basin and the New York area, for example, as much as 900 metres of salt was laid down.

Only slowly did the sea make its way back again into the continental interior and in the Appalachian area it seems to have been

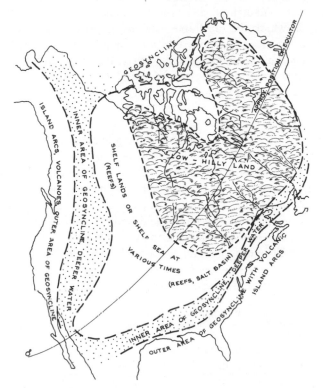

Figure 7-2   North America may have had a geography rather like this in early Palaeozoic times. On several occasions, however, much, if not all, of the low ground of the Canadian shield became submerged.

a sandy sea with many shellfish but not so many corals as previously. After this slow start the flooding began to quicken so that in middle Devonian time a few million years later it reached across the interior around the Canadian Shield. Only the Transcontinental Arch, the Ozark Dome and other minor regions were not covered. At this stage several interesting developments took place. East of the Transcontinental Arch the shallow, clear sea favoured many kinds of life and in it innumerable corals and other organisms set up house and built small patches of reefs. To the west, however, the shallow waters spread over an area that began to warp gently into one of the most remarkable of shelf basins, the Williston Basin. Here in the shallow waters coral reefs and atolls up to 65 kilometres across were soon sprouting on several parts of the basin floor. Between the reefs and the shore of the Canadian Shield or the land masses to the west were giant lagoons in which the inevitable salt and gypsum were precipitated. For as far as the eye could reach this would have been a seascape of blue water with foaming patches where the reefs and shoals were scarcely concealed, and of gleaming salt flats and beaches around the coastal lagoons. Blue seas and blue skies with few clouds to bring rain to the arid lands nearby persisted for millions of years. Beneath the surface of the water the floor of the Williston Basin was slowly sinking, but the growth of coral and the glistening layers of salt kept pace with this gentle subsidence. Between the reefs were gardens of algae and schools of small sea animals busy in the clear warm water. All in all, it may have resembled parts of the Red Sea today except that there were no cliffed highlands round about, only the low promontories of the Canadian Shield and the arches to the west. A closer look between the coral fronds would reveal a different assortment of small marine animals from those of the shellfish, crustacea, starfish and others we know in present reefs.

So abundant was life that there was a constant rain of dead creatures and decomposing matter falling to the sea floor. Much of it became trapped and buried in the sediments there. With the passage of time and the accumulation of a great weight of strata above these layers the organic materials experienced a great change; the original biochemical constituents were reorganised into the complex hydrocarbon molecules of petroleum. The deposits of the Williston Sea gave rise to oil and gas in huge quantities that were preserved in the porous reef rocks and limestones close at hand. From the sunshine of the Devonian period energy was trapped by countless living things, stored in the organic chemistry of their bodies and with them entombed in the rocks. The reefs were sealed over, so to speak, by layers of impermeable shale so that as the chemical transformations took place the resulting oil was kept bottled in, unable to escape upwards as it has done in so many other places. In 1947 the drillers reached these buried ancient reefs and the Canadian oil boom began. Since then the lines of the Devonian reefs have been probed by every possible means for the great quantities of oil that many of them contain. Most of the reefs are only a 100 metres or so thick, a rare few reach 400 metres.

After this long period of coral seas and shallows around so much of the Canadian Shield, it might be expected that a wide withdrawal of the waters would have occurred. No such thing happened. Early in Carboniferous time the continent seems to have slid quietly under the

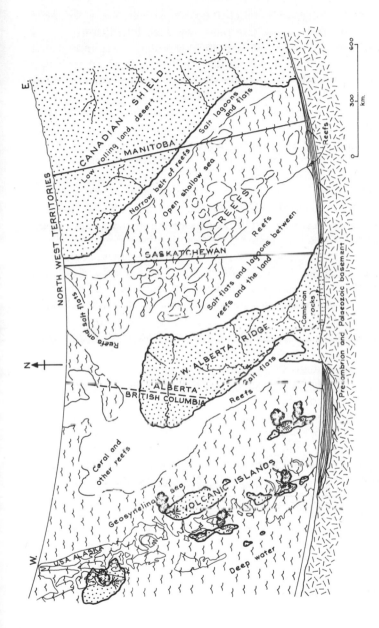

Figure 7–3   The Williston Basin was a large arm of the sea stretching down from the region of the North-west Territories across the prairie provinces of Canada into the northern states of the U.S.A. in middle Palaeozoic times. In these shallow tropical waters life was prolific and along the coasts there may have been extensive flats and tidal reaches where salt and similar materials were left by the evaporating waters. It was in fact one of the largest salt basins ever to have existed.

waves to an extent scarcely matched before or since. For a very brief period it seems that everywhere there was such little disturbance of the sea by currents or waves and so little water or sediment being washed in from the land that from the Cordilleran area to Appalachia there was stagnation. The corals died off, the shellfish and other invertebrates too came to an end. Somehow this once-thronged scene became an expanse of dead, still water and black mud, bereft of most recognisable living things. Exactly what happened is a mystery, a thin black shroud of shale hangs over the problem.

This almost lifeless episode passed by and organisms began to invade the region once more. Slowly the waters became populous again. Beds of shells, limestone sands and muds spread across the sea floor, and from the North-west Territories of Canada to Mexico and from the Pacific Ocean to east of the Mississippi there was once again a shallow sea. Known as the Madison Sea, this was a body of water in which the familiar production of lime carbonate sediments continued under warm, if not tropical conditions. The sea was clear and turbulent, the waves and currents kept the shells and sea lilies that grew in such profusion moving to and fro on the bottom. What accumulated on the bottom was the wave-worn debris of their skeletons, a sort of shell-sand. It grew to many hundreds of metres in thickness, the pulverised remains of astronomic millions of shells spread as a layer, many thousands of cubic kilometres in volume.

This indeed was the last of the great Palaeozoic floodings of the North American continent. There were to be gentle earth movements throughout the remaining periods of the Palaeozoic that warped the surface of the basement and its overlying sheets of sediment. Much of the Canadian Shield and the centre of the continent were raised, local basins were filled with silt and coarser sediment. The seas drained away to the south and north and in the geosynclinal regions more pronounced changes were taking place.

So at the end of some 300 million years the continent was due for a change such as it had not experienced before. The shallow seas ebbing and flowing between the shield and the geosynclines to east and west came to an end. Their record is preserved in the flat-lying, layer-cake, geology of much of the Mid-west, the fossiliferous limestones and shales of five geological periods.

Coal Swamps

Not all the environments that took shape on the down-warping margins of the shields were part of the marine realm. The name Carboniferous was given with good reason to the period following the Devonian and it is almost synonymous with the words coal swamps. It was a period during which the plant kingdom reached an unprecedented luxuriance. A combination of physiographic, climatic and biological factors made the opportunity for forests to spring up along the low-lying wet and swampy margins of several continents and across wide areas of the continental interiors between the uplands or mountain ranges. Ferns and other groups of rather primitive fern- and tree-like plants grew to enormous sizes and in great abundance in the hot, humid, swampy areas. They were moisture-loving, but they could not tolerate salt water, so that when sea level rose, as it did from time to time in the Carboniferous and Permian periods, the forests of the coastal marshes were killed off. When the sea withdrew or piles of sediment raised the level of the ground, new spreads of the forest could take root.

Dead and rotting plants lay upon the forest floor and despite the ceaseless work of algae, bacteria, fungi, and the other destroyers of dead tissue, their remains persisted as a layer that grew little by little each year. The process happens today in peat bogs, swamps and other places where the acid waters prevent the complete decay of the dead plant matter. A tough spongy layer of blackened debris, leaves, stems, roots and all is produced. This is peat and it is a complex of carbonaceous material which will, after the water has been removed, burn.

Under the weight of an overlying spread of mud or sand, and given plenty of time, peat becomes compressed into a flaking tough brownish substance, lignite, which burns much as peat does. Buried deeper and for longer time lignite loses more water, and other volatile constituents; it becomes compact, denser, black and to a degree shiny. At this stage it approaches the composition and appearance of coal. In fact coal is produced in exactly this manner by the continued compaction of a layer of vegetable matter to about one-fifteenth of its original thickness. The more volatiles that are squeezed out by pressure and heat, the higher is the proportion of carbon that remains behind and the harder, brighter (and more anthracitic) is the coal.

Carboniferous coal seams occur together with shales, clay, sand-stones and very rare limestones in what are called coal measures. 'Measures' is the old miners' term for the varied rocks that occur in a repeated and recognisable order between individual coal seams. The seams themselves may be a few centimetres or even a couple of metres thick: the measures between the seams are commonly many metres thick. In most coalfields the miners had, and have, little use for the rocks separating the seams, but early in the history of coal mining they found that these rocks lie in a definite order between one seam and the next above. The order results from a series of changes during the life of the coal swamp, events that were repeated in the same order many times. We believe the events formed what has been called a cycle of sedimentation and it was rather as follows. The coal plants established themselves on a soil from which they drew the mineral salts needed for their enormous growth, so for perhaps many thousands of years the forest flourished and the peat layer grew. Then, comparatively rapidly, sea level rose and the forest was killed off. Trees fell, rotted and collapsed into the stagnant black mud that everywhere smothered the old peaty ground. As the sea crept higher the water cleared a little: marine creatures and the tidal bank little lampshell, *Lingula*, came into the lagoons and reaches in large numbers. From the uplands periodic torrents of mud, sand and gravel spread into the sea, stifling the communities of shell-fish and other sea creatures. Water courses, lakes, bayous, sand banks and mud flats formed and dispersed over the years while the land-derived sediments kept the sea in check. After heavy rainfall the streams were charged with sand and clay, masses of broken and decaying vegetation and driftwood were swept along towards the sea. There were log-jams, river banks burst and splays of sand deluged onto the backwater flats and marshes. Locally fish and amphibia in the waters were suffocated and buried in the silt. What had been a forest swamp and later a lagoon or tidal reach became a delta. The intricate network of streamways was constantly changing but the thickness of silt, mud and sand slowly grew, compressing the buried peaty layers unceasingly.

Wherever the opportunity offered, the coal swamp plants came back to colonise the ground. Every so often they were successful enough to establish themselves from horizon to horizon. Tall plants formed a canopy above smaller shrubby and reedy communities and on the ground itself lichens, liverworts, fungi and others were in possession.

Long as this forest may have lasted, it too succumbed as had the previous vegetation. Sea level rose again, or the land itself sank and the cycle of events was set in motion once more. Such very large areas were affected when these changes took place that it seems likely that world-wide sea level may have been involved. Several things might have been responsible for these oscillations, either the actual amount of sea water changed or there were changes in the volume that the ocean basins could hold. Possibly there may have been repetitive earth movements that raised or lowered the land surface at regular intervals. Climate may have followed a cyclic or oscillatory pattern with wet periods and dry periods. Perhaps it was all of these in combination to some extent that affected such widely separated areas as Kansas and the Appalachians, Britain and Western Europe, the Don Basin in the U.S.S.R. and parts of the southern continents.

We know that these cycles, or *cyclothems* as they are usually called, vary in the proportion of marine and non-marine sediments that they contain from place to place. By measuring these proportions, by finding traces of current directions in the old river sediments and by recording the variations in thickness of the cyclothems we can suggest where land and sea lay. In many parts of the world cyclothem is piled on cyclothem until the total runs in scores. The cause of this is still a mystery but it seems to have faded away in late Carboniferous and early Permian times. These were indeed periods of much mountain-building and volcanic activity, and throughout this time the world climate seems to have been changing. In the south polar region a series of continental glaciations was taking place, but the lowering of sea level in response to the removal of water to snow and ice is hardly likely to have fluctuated so often as to be directly the cause of the coal swamp cyclothems. But as Permian time passed a world-wide increasing aridity and withdrawal of sea from the margins of the continents made itself felt. If this were due to the actual movement of the continents relative to one another and to the changing shape of the ocean and its floor, just possibly the cyclothems too were the result of such movement.

At all events, cyclothem deposition came to an end with the upsurge of the Appalachian earth movements in North America and the Hercynian (or Variscan) orogeny in Europe. For a while, most of the interior of North America had become land. The great day of the swamp forest was over. The continental interior was now to remain more or less free of the sea until late in the Mesozoic. Its Precambrian basement was effectively buried several thousand metres beneath the surface around the Canadian Shield. As the burial had been taking place the old shield had been eroded to provide much of the sediment. Subsidence had been far from uniform everywhere, and the way in which subsidence occurred had largely controlled the extent of land and sea. A similar pattern may emerge from the other continents as our knowledge of them improves.

Although the 'layer-cake' geology of the once sea-flooded areas of the continents is in fact far more complicated than we have been able to describe here, it does bear out the contention that the continental shelves and platforms do record in broad terms the major events in earth history. The great transgressions and regressions of the seas have left their mark and these can locally be related to movements of shield areas or mobile belts. Life, too, has left a record of the growth, evolution, and decline of marine 'brackish water' and swamp communities, and it has bequeathed the coal and petroleum deposits that have influenced the development of human activity.

The Karroo Basin

Here we are concerned with a different kind of geological story, one concerned with an upland continental shelf area largely isolated from the sea and which nevertheless subsided beneath layer upon layer of sedimentary deposits. Only at the eastern edge of the continent did it meet the sea, elsewhere it seems to have been bounded by hills and escarpments. This major feature in African geology is known as the Karroo Basin after the Karroo area of South Africa. At least 7,000 metres of continental sediments, many containing abundant fossil plants and animals, were deposited here between late Carboniferous and mid-Triassic times.

Much of southern Africa and the other southern continents was capped by an ice sheet of gigantic proportions in the late Carboniferous period. The evidence for it is abundant, indisputable and spectacular. As much as 400 metres of glacial boulder or moraine beds, sands and gravels lie directly upon the Precambrian basement in the Karroo Basin itself. But it is after this glacial period that the events in the basin acquire a special interest for us. The glacial rocks are overlain by the mudstones and limestones of lakes and lagoons, which contain the remains of a small swimming reptile, *Mesosaurus*, who turns up again in Brazil.

On top of these formations comes, rather unexpectedly, a coal measure sequence. The Ecca formations are about 1,800 metres in total thickness and contain many beds of thick coal. The flora of these coal swamps was quite different from those in the northern continents at the time. All kinds of plant fossils have been found including seeds and well-preserved, though petrified, fruits. Because many of the trees have growth rings in the trunks, because of the great abundance of leaves, and because of other anatomical peculiarities this *Glossopteris* flora is thought to have been deciduous, living in cool temperate swampy areas. Between glacial spells in the Carboniferous period the *Glossopteris* and other trees covered the land. In the undergrowth there were many animals no doubt, and among them were some of the first true reptiles. At this point in evolution the reptiles were quite novel and were by far the most advanced forms of animal life.

The Ecca sediments are quite clearly very like those in coal-bearing sequences elsewhere. The coals have their underclays and roof shales, and are continuous seams over huge areas, just as in other continents.

The Ecca formations were deposited during the Permian period, to be followed in the early Triassic by some of the most important strata to preserve vertebrate fossils. In the Beaufort beds, which reach over 3,000 metres in thickness in the south, there is a fabulous record of the reptiles. They seem to have reached a point at which suddenly they

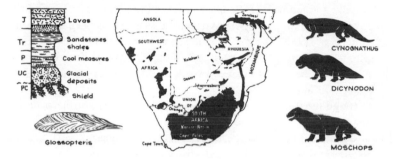

Figure 7–4　The Karroo Basin preserves a remarkable sequence of glacial rocks, coal-bearing rocks and red sandstones. The fossils include the famous *Glossopteris* 'ferns' and mammal-like reptiles in countless thousands. All in all, this is one of the world's most remarkable accumulations of sedimentary rocks on the surface of an ancient continent.

could branch out into several new modes of life. Some were the size of a large pig or dog and the biggest was as large as a rhinoceros. The earliest of the Beaufort beds has nearly fifty different kinds of reptiles among its fossils. The sediments were deposited in stream courses, river flats and plains, in a rather warm climate that had arid spells when the bodies of water dried up. This was very hard on the fish and amphibia in the basin but the reptiles seem to have found it most congenial. Literally millions of reptiles must be preserved in these rocks. What the total numbers of living animals may have been is hard to guess. Perhaps they lived in herds and packs wandering along the water courses, following trails and migration routes as African animals do today.

A remarkable feature of this fauna, which we shall again note when considering continental drift, is that the reptiles seem to have had close relatives in Russia and in South America, Australia, India and Antarctica. Although the Russian and African plants were so different the animals managed to spread to both areas without obvious difficulty.

Last of all the Karroo formations are the Stormberg. These sedimentary rocks and the fossils they contain suggest that the climate was becoming increasingly warm and dry. *Glossopteris* died out and shortly afterwards most of the other plants did too. Among the last of the reptiles to be found in the area were early, though not large, dinosaurs. The curtain to be drawn over this wide scene was eventually a volcanic one. Basaltic lavas poured from fissures and rifts to cover a large part of South Africa. These Drakensberg lavas reach as much as 1,000 metres in thickness, flow upon flow marking eruptions continuing to the very end of the Triassic period. Even after long erosion there is still 20,000 square kilometres or more of the basalt remaining.

Throughout this spell of 85–90 million years the southern part of Africa and parts of the other southern continents had become covered with deposits like those described here. The years had seen the *Glossopteris* forests bloom and die and the reptiles flourish and scamper away. The sea rarely and only briefly crept into what are now the interiors of the continents during this time but in the end floods of basalt sealed over these parts of Africa as they did in parts of other continents. Nowhere and at no time in this part of the evolution of African geology were there violent or long-continued earth movements. Even the ultimate eruptions were relatively quiet affairs. The underlying basement may have been fractured to let the lava escape but it behaved throughout this period in the most passive manner.

Those days, however, have been followed by periods of gentle uplift, and if there were further sheets of sediment laid down they have since eroded away. South Africa has a long Mesozoic and Cainozoic history of land rising slowly, just as it had subsided before. Erosion has not been so very vigorous, largely because the region is rather a dry one. The Great Karroo today lies between 500 and 1,000 metres above sea level between the great escarpment of basalts and the folded rocks to the south, a dry area known to be good sheep country. To what do we owe the accumulations of the Ecca coal measures? Were the conditions for the spread and later sudden death of the *Glossopteris* forests related to the kinds of crustal activity that Eurasia or North

America were then undergoing? There is no sign of nearby or recent mountain-building when the Ecca forests subsided beneath floods of mud. Perhaps short pulses of gentle uplift over the continent beyond the basin compensated for the sagging within it, and the sediment-laden streams flowed down the new slopes with increased vigour. After the release of the Stormberg basalts this movement was succeeded by long and slow uplift which may still be going on.

Ever Upwards

So in the course of the last few hundred million years the continents have been flooded to a greater or lesser extent. At one time some regions emerge from the shallow seas, at another it is the turn of other areas. Geologists have not unnaturally to ask if there has been a pattern to these dousings or if a particular trend has become apparent. Some European geologists, at least, believe that there has been a progressive decline in the proportion of the continents covered by the sea since the beginning of our Phanerozoic eon.

Could the sea have been diminishing in volume? This seems highly unlikely. In fact volcanic additions may have increased it very slightly. Perhaps the opposite has taken place and the earth itself increased in volume. Thus the waters would only be able to cover smaller and smaller areas of the continents as time went on. Cretaceous marine rocks offer indications of a sharp reverse to this general trend, covering more of the continents than had been under water since the Carboniferous period.

A close look at earth's geochemical and geophysical attributes and its crustal structures and behaviour lend no support to the last suggestion. We see no real reason to believe that the total mass of the earth has expanded or decreased to any appreciable degree during geological history.

Oxford geologist Tony Hallam has suggested that simple isostatic uplift as a result of a slight thickening of the continents during the Phanerozoic eon would give the effect of reducing the areas of marine inundation. He points out something that has been implicit in our account of how sheets of sediments eroded from the old crystalline shields have come to rest on the shelves roundabout. Erosion of a shield produced isostatic uplift, exactly as an iceberg rises when its top melts away, and erosion has been going on for many million years. But obviously this erosion and uplift cannot persist indefinitely because, in the end, the shields would achieve isostatic stability with low relief and would be relatively thin. The strange fact here is that the crust beneath the continental shields is not particularly thin; quite the reverse, for many of them are thicker than the average for the whole continent. As we noticed earlier, large areas of the outcropping rocks are of once deeply buried materials and of greenstones. The greenstone belts, you will recall, were thought of as forming when the crust was not particularly thick. In those early Precambrian times crustal thicknesses may only have been 12–15 kilometres. From Africa and other shields there seems, then, to be good evidence that some sort of addition to the underside of the continental sialic slab has been taking place over the last few hundred million years, if not longer.

There is another line of argument that suggests that the lower surfaces of the continental slabs or plates may be reinforced by additional material from time to time. In Chapter IX we shall see how portions of the crust may be carried down into the mantle under the edges of the continents. During this rather drastic process enormous heat is generated and magmas are produced from the descending chunks of ocean floor. Some of this new magma works its way to the surface and causes explosive andesitic volcanoes, some is held below to be intruded as granite and some may spread over the undersurface of the continent. It is this last fraction, seeping, as it were, in from the edge towards the centre of the continental plate that has been so important. The thickening it has produced has kept the shields rising slowly but surely for so long.

Estimating how much the shields have risen since early Palaeozoic time is a difficult matter, but 300 metres cannot be far wrong. Thickening the continental crust by twice this amount would most probably secure the actual uplift. It is in fact a very small amount—an increase of 1 millimetre in 10,000 years. In itself that is less than 2 per cent of the present average thickness of the continents added since the dawn of the Cambrian period.

The Karroo Basin, nevertheless, represents a long time of regional subsidence of the shield, rather than uplift. Could this not be a period corresponding to a cessation or slowing of the additive process under the continent? Quite possibly; and it is terminated by a sudden resurgence of crustal activity, but in another form—the great floods of basalt.

Perhaps the puzzling development of arches and domes and basins and troughs upon the shelves is all a reflection of isostatic change produced by uneven increments of light sialic material to the continent from beneath. This explanation could hold for those Palaeozoic features seen in the stable interiors of North America and for some, at least, in Africa.

But if this is a long-continued, slow-moving process, undeniably likely to lead to the prolonged uplift of the shields and the transfer of material as sediment to the shelves, there is likely to be another agency, more rapid and ephemeral, causing the major spreads of the sea over the continents from time to time. The equally rapid and marked withdrawal of the seas in say the Permian and Triassic and the early Cainozoic times has to be accounted for. Here, too, the crustal machinery, and especially the relationships of continents and oceanic ridges, is important. The business is rather more involved and an explanation is best delayed until Chapter IX. No changes in the volume of the sea are really necessary and no obscure glaciations are needed to steal the water. Once again, we can look to the behaviour of the deep crust to explain the changes taking place on its surface.

CHAPTER VIII

Continental Drift

BY THE end of the nineteenth century it was obvious that to make sense of the maps drawn for each geological period on each continent, there were some rather substantial problems to be solved. Fossils, both plant and animal, were known to be distributed in ways that suggest land connections between areas now widely separated. 'Land bridges' was the term used for these convenient regions which stretched across the oceans to link, for example, Africa and South America, or Europe and North America. They look rather odd on world maps and even their authors must have wondered how such major geographical features could have risen and subsided so vigorously. When a land bridge rose to connect, say, North America and Eurasia, a lot of ocean water would have to be displaced. Did it flood the continental margins round about? When a bridge sank did the ocean elsewhere become shallower or deeper? Land bridges may have seemed to offer some help to the palaeontologist with his eye on the possible migrations of ancient floras and faunas but in these other respects they posed some knotty problems. And what had provided the energy and means by which the land bridges had risen and sunken?

To some of these questions at least an answer was provided by a German geophysicist and meteorologist, Alfred Wegener, in 1912. As he saw it, the continents themselves had moved, split from a single parent land mass and drifted to the positions on the face of the earth where now we find them. Wegener gave a full account of the evidence which led him to the idea in his book *Die Entstehung der Kontinente und Ozeane* in 1915, but he acknowledged that the American, F. B. Taylor, had put forward something like it in 1908. Taylor had been very struck by the apparent fit of the coasts of Africa and South America. He proposed that the rifting and displacement of the continents had caused the circum-Pacific ring to form and the Tethyan ranges to be pushed up. This was perhaps not too difficult a pill for the geologists of the time to contemplate swallowing, but Taylor suggested that the movement had been a tidal slithering of parts of the crust caused when the moon was supposedly captured from outer space in the Cretaceous period. The idea did not appeal to the scientific public.

Also impressed with the possible reassembly of continents into a single land mass and their flight under tidal influence was another American, H. H. Baker. In 1911 he depicted a tidy clump of continents that was to split when Venus briefly approached the orbit of the earth in the Cainozoic. He thought, as George Darwin had done before him, that the moon could have been pulled from the area of the

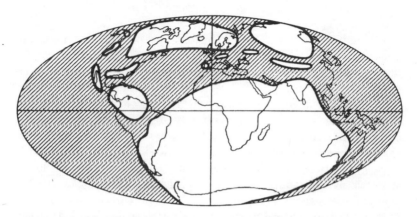

Figure 8-1   A view of Carboniferous geography which does not invoke continental drift gives the world picture shown here. It is a very different map from that drawn by Wegener and it involves 'land bridges' joining North America and Europe and the southern continents. Wegener found it incompatible with the evidence he had and did not see how the 'bridges' could founder without a trace.
(From The Origin of Continents and Oceans, *by courtesy of Methuen & Co.)*

Pacific basin. Both the lunar birth and the drift of the continents could be the result of Venus trespassing close to the earth in Baker's theory.

Leaving aside all temptations to invoke the planets to disrupt the earth, Alfred Wegener concentrated with characteristic firmness of purpose upon the tangible geological evidence. He felt that our earth had forces enough to achieve the changes that he could demonstrate to have taken place.

Just as many other new or revolutionary ideas in science have been given a rather lukewarm reception, Wegener's book met with incredulous or sceptical reaction. Some geologists were openly hostile and may have felt that a meteorologist had no business flirting with such unorthodox geological concepts. Nevertheless Wegener was not dismayed: he was nothing if not thorough and in subsequent editions of his book in 1920, 1922 and 1929 he was able to add more data and cogent argument in favour of his hypothesis. Translations of the third edition into English, French, Spanish and Russian helped bring it to a wider public and there is no doubt that Wegener was heartened by some of the views expressed by geologists outside his own country. The fourth edition was translated into English only in 1966, significantly after the new bandwagon had begun to roll.

Alfred Wegener lost his life while a member of an expedition to Greenland in 1930 but he left behind one of the most important concepts to have appeared in geology in the twentieth century. The final edition of his book is a masterly summary of the evidence for the hypothesis. Wegener put palaeontological, stratigraphical and structural arguments together in a way which is easy to follow, building up a picture for each continent and showing how the resemblances between the continents can be used as a basis for reconstructing the existence of a single parental continental mass.

From South Africa in the 1920s and 1930s came a steady flow of research by Alexander du Toit, a geologist who was in 1937 to champion Wegener's cause in a splendid book *Our Wandering Continents*. Du Toit's careful studies of South African and South American geology put flesh on some of the bones that Wegener had assembled. There were other studies in both northern and southern hemispheres which continued to add weight to the idea of continental drift, but the means by which drifting was achieved were far from clear. Arthur Holmes, at the University of Durham, had welcomed Wegener's ideas on drift and offered the suggestion that convection currents in the earth's mantle could provide the means of moving the continents from place to place. Holmes was a pioneer in the use of radioactivity to measure geological time and saw in radioactive heat a continuing source of energy within the earth, seeking outlet from time to time by convection. Here, perhaps, he had a continent-shifting force.

Not every geologist or geophysicist was ready to subscribe to Holmes' ideas any more than to those of Wegener. Wegener believed that the forces that pushed or pulled the continents were the same as those that produced the great mountain ranges. As we have already seen, the prime force in bringing about the structures in mountain ranges is one of lateral compression.

Today the great majority of geologists accept the idea of con-

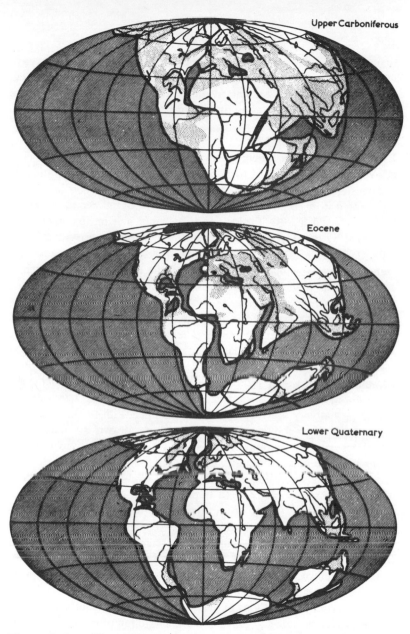

Upper Carboniferous

Eocene

Lower Quaternary

Figure 8–2   Alfred Wegener's reconstruction of Late Carboniferous geography gave the scientific public a completely new view of earth behaviour (compare with 8–1), and it aroused much criticism. However, it is now generally acknowledged that Wegener's was a more plausible interpretation than that of his critics.

(From The Origin of Continents and Oceans, *by courtesy of Methuen & Co.)*

tinental drift and are relatively content that it explains the odd ways in which certain rocks, fossils and geological structures are distributed. Here are some of the arguments Wegener used sixty years ago and some of the additional discoveries that have been made supporting them.

Evidence in the South

Ever since the days when the first maps of the world were being drawn, the striking 'fit' of the opposing coastlines of South America and western Africa has roused comment. Francis Bacon is said to have noted it in 1620, which is certainly early by all accounts, and it had impressed F. B. Taylor, and Wegener began his geological argument by considering it. The opposing coastlines across the South Atlantic look as though they could be fitted back together with few 'gaps' or 'overlaps' between them. When the edges of the continental shelves rather than today's coastlines are considered the fit is much better. A Tasmanian geologist, Warren Carey, using a large globe recently made cut-outs of the submarine contours to test their fit and found it looked very convincing. The whole business of a fit was re-examined a few years later by Sir Edward Bullard and his research students, J. E. Everett and A. G. Smith, with the aid of a computer, and it was found that the best fit was obtained by using the 500 fathom contour, a theoretical edge to the continent which in fact lies about halfway down the continental slope. With further aid from the computer a convincing map could be drawn to match the coastlines with a minimum of misfit and when the geology was added to the map the fit became more convincing than ever. Although there are three areas where the outlines overlap, we have good grounds to discount them. The Niger delta impinges on north-east Brazil, but it seems to be a thick wedge of sediment no more than 50 million years old and grew only after drift had begun. Nearer to South Africa there are islands and a submarine ridge which seem to block a good fit, but these too have grown in place after drift had started.

Over much of eastern South America and Africa the Precambrian rocks lie at the surface. Their internal structures are complex and were acquired over many millions of years on occasions when the rock now at the surface was deeply buried. Wegener's champion and advocate, the South African, du Toit, thought that the match of such belts was good argument for their previous continuity. When radiometric dating tests were applied thirty' years or more later it could be seen that the orogenic belts not only matched in their trends and fabrics but also in their ages.

Figure 8–3 The fit of the continents around the Atlantic made by Sir Edward Bullard and his assistants at Cambridge, using a computer to match the 500 fathom (approximately 1000 metre) contours. The overlaps are shown in black and the gaps stippled. It is thought that the actual edges of the continents lie at about 1000 fathoms but the fit at the 500 fathom mark is rather better than at the greater depth.

When the radiometric ages of the Eburnian and Pan-African structural provinces were established at 2,000 and 550 million years ago respectively it was found that the provinces met at Accra on the west coast of Africa. The sharp boundary between them trends out to sea to the south-west. American and Brazilian geologists, suspecting that they might find the drifted counterpart of these provinces in eastern Brazil, carried out radiometric dating analyses over a wide stretch of the north-east coast of South America. North of the city of Sao Luis in Brazil the Precambrian is 2,000 million years old: to the south and east it is 550 million years old. The dividing line falls exactly where it was expected on the map when Brazil and Africa are reunited. It all seems to be good evidence to suggest that the continents were joined together 550 million years ago but had begun to separate before 50 million years ago.

Perhaps the next piece of geological evidence is even more convincing and is more useful since it involves all the southern continents, Australia, Antarctica and India as well as Africa and South America. It is of course the great series of rocks laid down as sediments under glacial conditions during Permo-Carboniferous times, 350–200 million years ago. They go by different names in each continent, as we have seen, but as the Dwyka Series in South Africa they are perhaps best known. In Africa the Dwyka Series occurs over much of the country between the southern cape and the equator. It is a remarkable thick series of shales, sandstones and pebbly beds, blanketing the much older rocks beneath. The Dwyka rocks resemble in every way the deposits left by glaciers and their run-off streams in the northern hemisphere during the Pleistocene ice age. Moreover there are grooved, polished and striated surfaces below the Dwyka, exactly like those under modern glacial sediments, and there are large blocks far from their original sources just as there are glacially travelled erratic blocks today. In short, the Dwyka must be a glacially deposited series of sediments, and must have accumulated either in a polar region or in high mountains. Nothing like the Dwyka was being formed in the northern continents at that time.

So extensive, thick and uniform are the Dwyka glacial beds that they could not have been formed in mountains, but they could have originated around the Permo-Carboniferous south pole. In many places they are 600 metres thick, several times the average thickness of the ice age deposits in Europe, and so they must have been derived from a very large source area undergoing erosion for a long time. A continental polar region answers the demands nicely with glaciers carrying debris off radially from around the pole. When the signs indicating the directions in which the Dwyka ice was moving are read the flow of those ancient ice sheets and glaciers can be plotted. In general it does seem to be radially outwards from a pole situated on the coast of present day Antarctica, and in South Africa it flowed south and west, towards South America. The deposits are meagre in south-west Africa, as though the region were being heavily eroded and not much smothered by newly produced debris. The weathered rock and soil was being pushed off westwards. Some of the isolated and 'foreign' boulders in the rocks in Brazil seem to match material in African boulder deposits and some can be matched to bedrock in Africa.

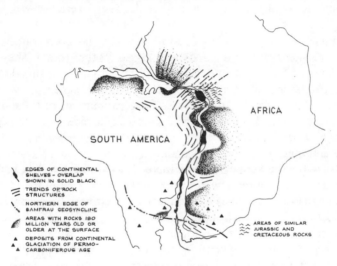

Figure 8–4 Like pieces of a jigsaw, similar geological features can be matched upon the opposing coastlines of Africa and South America. This kind of matching can be done between other continents but it is by no means everywhere so convincing and there are alternative fittings.

Of course these facts don't prove that South America and Africa were once of a piece but they do suggest that no great Atlantic separated the two.

Similarly, the Ecca Series with its coals and distinctive *Glossopteris* flora in South Africa is matched by rocks and fossils of an identical nature in the other southern continents. Elsewhere the Permo-Carboniferous landscapes were clothed in quite different kinds of vegetation, indicative of hot or tropical conditions. We find it hard to imagine the *Glossopteris* plants spreading to continents separated by oceans. The *Glossopteris* seeds were too heavy to be blown so far by the wind.

A related field of evidence that seems to cry out to be noticed concerns the Permo-Triassic Karroo and other rocks of the southern continents and the distinctive vertebrate fossils found in them. In Chapter VII we suggested how the rocks were at least in part deposited as sediments on a land surface or in shallow lakes and rivers. Wandering over the Permo-Triassic countryside were different kinds of mammal-like reptiles that did not survive the Triassic period. It is really a moot point whether these animals should be classified among the reptiles or with the mammals. There were herbivores and carnivores, in size ranging from a rabbit to a large dog. As in mammals the legs were 'pulled in' beneath the body rather than sprawled sideways and the body was raised off the ground. They were well adapted for life on land, seeming to be perhaps best suited for the open plains or veldt; they were certainly not designed to swim the seas. And yet several of them occur in more than one of the southern continents; even Antarctica has its terrestrial mammal-like reptiles from the Triassic rocks.

In addition to the similarities shown by the Permo-Triassic reptile faunas of the southern continents there exists also a close link between them and fossils of the Dvina Series in the Perm area of the U.S.S.R. So many close similarities exist that there must have been continuous land between Perm and South Africa.

Several accounts of the evidence for the close proximity of South Africa and South America in the Permian period point out the presence of a small aquatic reptile, *Mesosaurus,* in rocks of this age in both continents. If *Mesosaurus* swam in the sea he could no doubt have crossed a small ocean or seaway, but if this was a freshwater-loving reptile it would be remarkable that it occurred in widely separated continents. Despite our inclination to wish a freshwater habitat upon *Mesosaurus,* the fact remains that it may have been a marine beast. Even if *Mesosaurus* never had deliberately put foot or flipper into the sea, the presence of the fossils in South America and South Africa is no absolute proof of the continuity of these regions in Permian times.

Up to this point in the geological record of these two continents much of what has occupied our attention has suggested that they were once closely placed together. What comes next hints at their growing separation. Along the eastern seaboard of Brazil and the west coast of Africa are several thick deposits of late Jurassic and early Cretaceous date. The sedimentary characters and fossils in these rocks indicate bodies of fresh water. Amongst the fossils both in Brazil and in West Africa are ostracods (tiny active creatures with a bivalve shell) that could live only in fresh water.

Above the freshwater deposits are beds of salts left from the evaporation of sea water. The old freshwater lake region had been invaded by the sea and then had dried up. In successively higher beds (and therefore younger beds) above these salt deposits there are more and more differences to be found between those of Africa and those of Brazil. Each area seems to have developed independently from then on. We can date the salt deposits as about 100 million years old and that may have been the time when the sea began to creep in between the uplands of Africa and those of South America.

And Evidence from the North
While the 'fit' of Africa and South America is good and the geology within the continents tends to 'match' across the Atlantic, the 'fit' of the continents about the North Atlantic is not so obvious. In many ways, too, the geology of the countries bordering the ocean there is more complicated. Palaeozoic and some Precambrian rocks make up the greater part of the complex geological picture. Many of them were known in detail in Europe and North America in Wegener's day and the similarities aroused much comment. Today we recognise that there are parts of the Precambrian in the British Isles which are of the same age and in the same condition as Precambrian rocks in Greenland. The overlying Palaeozoic rocks in Western Europe and eastern North America and Greenland show many close similarities and contain similar fossils. They seem to have been deposited under identical geosynclinal conditions. Moreover, they contain the same kinds of fossils. Earth movements throughout early Palaeozoic times occurred frequently in both areas and reached a climax in the Devonian period. The great upheaval that resulted then left matching imprints on the rock structures now to be seen in East Greenland, Norway, Britain and eastern Canada.

Known as the Caledonian orogeny, this climax was accompanied by the intrusion of granites and widespread alteration of the old geosynclinal sediments. Resting upon the eroded stumps of the Caledonian rocks are the Old Red Sandstone formations. Boulder and pebble beds, sands and clays derived from the underlying formations, these beds also contain the remains of strange and armoured freshwater fish.

In Wegener's closure of the Atlantic Ocean, and in other versions too, the areas of Caledonian rocks and the Old Red Sandstone outcrops all seem to group together very naturally in a narrow band running from Spitsbergen to the British Isles and on to Newfoundland and mainland North America as far south as New York. Here clearly we have the remains of an old mobile belt which seems to have been partly split longitudinally down the middle.

Remarkable as this is, it is paralleled by another such belt involving principally Devonian, Carboniferous and Permian formations from Alabama to Newfoundland in eastern North America, Britain, mainland Europe and coastal north-west Africa. Rocks of earlier ages are locally present but the orogeny, known as the Variscan or Hercynian, took place there between 370 and 290 million years ago. The rocks, the fossils, and the kinds and date of the earth movements in these regions leave no doubt that this was another geosyncline-like belt.

Where in the north there had been sea in early Palaeozoic days there was land, later in that era, and in the southern regions, where open water existed until late in the Palaeozoic, there was similarly an upland region. If there had ever been a Palaeozoic proto-Atlantic it would seem to have been closed up by about 290 million years back.

Here then is evidence for a double drift. The closing of a proto-Atlantic by Permian time has been followed by its opening in Mesozoic and Cainozoic time. To Wegener and others the late Mesozoic and the Cainozoic volcanic activity that could be found around the margins of this ocean were evidence of the rupturing that took place.

Magnetic Play-back

Despite the powerful advocacy of Arthur Holmes and other able geologists, the theory of continental drift had by mid-century not been accepted generally. It seemed to be a too-glib arrangement of circumstantial evidence in the eyes of many geologists and geophysicists alike. A change of heart came when palaeomagnetism began to offer evidence of a new kind. In Chapter III the idea was introduced that the earth has a magnetic field similar to one that would be produced if a simple gigantic bar magnet were embedded within the planet roughly down the axis of rotation. The lines of magnetic force emerge in the southern hemisphere and curve round to re-enter the earth in the northern hemisphere. For each latitude the angle of emergence in the south (or re-entrance in the north) is fied and characteristic.

When igneous magmas crystallise or when sediments containing iron minerals come to rest they take on a weak magnetism related to the lines of force present in the earth's field at the time. It seems not unreasonable to assume that the earth's magnetic field has always resembled that of a bar magnet and that rocks in formation have always acquired a little of that magnetism. If the direction of that magnetism in the rocks can be found we can, in theory at least, suggest where the magnetic pole lay at that time. In most cases it is a position far from the present pole, but if the pole itself never wandered far the apparent discrepancy would indicate that the site of the sample had been moved. So perhaps we can reasonably deduce the latitude of the rock as it first formed and point to the probable position of the poles then. It sounds simple—indeed it was first suggested in the late 1920s—but it has taken great effort and skill to develop a means to give palaeolatitudes reliable to within 10 degrees. What the method does not tell us, of course, is the palaeolongitude.

Thousands of specimens of rocks have been analysed for their palaeomagnetism and it is an obvious test for geophysicists to apply to see if rocks have changed their latitudes since they were first formed. It is, in fact, a test which can reveal if displacements on a large scale have taken place. Just what the proponents of continental drift needed, the palaeomagnetic approach, despite some limitations and drawbacks, was applied with vigour to the business of verifying continental drift. When applied to rocks of the same age in, for example, Africa and South America, some interesting facts begin to emerge. It does seem that both continents were indeed within the south polar circle during the Permo-Carboniferous glaciations but by late Triassic time they had moved about 40 degrees equatorwards and that during this time they had separated by several hundred kilometres.

So here is an independent line of investigation corroborating the drift hypothesis proposed on the basis of geological evidence. The impact of palaeomagnetic studies on the continental drift argument was immediate. Here were geophysicists producing evidence to support the idea. For many years it had seemed that all geophysicists could do was to find theoretical reasons why drift was ridiculous. One might even say that relations with geologists began to improve rapidly. Professor P. M. S. Blackett (later Lord Blackett), who might have been described as the 'Father of Palaeomagnetism', afterwards referred to the debate on continental drift at the 1950 meeting of the British Association for the Advancement of Science as 'the last great pitched battle between drifters and antidrifters'.

Gondwanaland—The Lost Continent

For decades before Wegener uttered his heresy there had been a growing body of geological evidence to suggest that all the southern continents were related, must have been connected somehow, and may even have been linked by land bridges. Eduard Suess, the indefatigable Austrian geologist of the late nineteenth century, had used the name Gondwanaland for the super-continent of which they were the last surviving fragments. Wegener kept the name for this continent too, but his idea of its size and shape were very different from those Suess had given. Wegener was most impressed by the Permo-Carboniferous glacial deposits in the present southern continents and his reconstruction of Gondwanaland relies heavily on this. Du Toit, working largely in South Africa, collected much more evidence to support the idea of the Gondwana super-continent. Support it, the new facts all seemed to do, but they were not proof.

Yet each part of this jigsaw puzzle is indispensable to the whole picture. There are plenty of geological features and fossils that suggest ways in which the pieces fitted together in the original whole. The problem is to find the best fit, the most likely arrangement, and the proof that this is how it was. South America and Africa seem to key together reasonably well. Now to add Arabia, Madagascar, India, Australia and New Guinea, and Antarctica.

Arabia does not present too much difficulty; the coastlines fit well. On the other hand, the Red Sea rift has been the scene of volcanoes for the last 25 million years or so, and volcanic rocks have been added to the edge of the African and Arabian continental edges. No one has yet found why.

We can add Madagascar to the picture by returning it to the East African coastline just south of the Horn of Africa. In this position it fits well geologically.

Australia and Antarctica also can be matched on general geological grounds with eastern Antarctica 'tucked in' to the southern edge of Australia and Tasmania. Much of the evidence from Antarctica has only been known for the last 10–15 years or so and of course it has been collected from the ice-free areas. In turn, Antarctica seems now to have held a place 'south-east' of Africa. Between Madagascar, Africa and Antarctica there is room for Ceylon and between Madagascar and Antarctica is a suitable resting place for India. Not unnaturally, this reassembly of these pieces of old continent leaves a few gaps between them and locally involves a few overlaps, but these can be accounted for in a variety of plausible

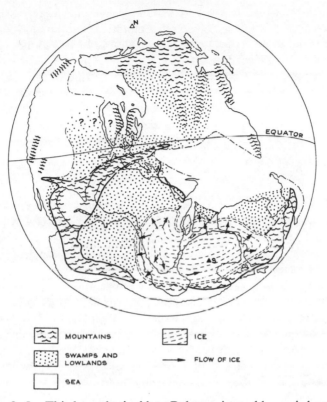

MOUNTAINS

SWAMPS AND LOWLANDS

SEA

ICE

FLOW OF ICE

Figure 8–5 This hypothetical late Palaeozoic world map is based on a number of different interpretations of the evidence and, as such, is a compromise, but it gives a rough indication of the extent of the south polar ice cap and the more conspicuous mountain ranges. The equatorial low areas were soon to dry out and become arid semideserts in Permian time. (After Dott and Batten.)

ways. The gaps really are remarkably few and, as reassembled in this manner, old Gondwanaland was a truly compact affair. Another recent idea about Gondwanaland is that it may have had a more or less unbroken mountain chain around it—an elevated rim. The effect of this upon Gondwanaland's climate would have been fairly conspicuous. We must await more work on this problem.

We saw how the geological indications were that Africa and South America began to part company in late Mesozoic times. It seems to

be much the same for the rest of Gondwanaland. Rocks as late as
Jurassic in Africa, Australia and Antarctica show many common
features but more recent formations have progressively more differences from continent to continent. The break-up seems here also to
have begun in the Cretaceous period. The parting of the continents
would not have been a violent cataclysmic affair, widespread though
it was. It seems to have proceeded as a steady unhurried and relatively
unbroken movement. At least so the evidence from the ocean floor
seems to tell us. As the continents separated, each tended to rotate
slightly in a sort of slow waltz.

Laurasia: A North Atlantic Alliance

Today's northern continents are only two in number—North America
and Eurasia—but there is also Greenland to accommodate. All in all,
these are some of the geologically best-known regions of the world
and the correspondence between eastern North American geology and
that of western Europe, as we have seen, has long been a matter for
comment. Not only is the 500 fathom contour fit proposed by Bullard
and others in Cambridge a good one, but the geology of both eastern
and western continental areas is easier to explain if their previous
juxtaposition can be proved. A highly successful international confer-
ence in Newfoundland in 1967 reviewed the matter in great detail. All
manner of geological and palaeontological evidence is found to be
available for use in the drift hypothesis.

A point that must not escape us is that to make the fit a good one
the North Atlantic and Greenland Ridges, Iceland and other islands
have to be removed. They are all made of rocks rather younger than
70 million years old. On the other hand, Rockall Bank between
Greenland and Britain is now known to be floored by a small area of
continental crust.

The 70 million year figure seems to give the date up till which
Laurasia was intact. The opening of the North Atlantic is thus a
phenomenon that has occupied little more than the Tertiary period.

Pangaea: The Ultimate

Not content with two super-continents, Gondwanaland and Laurasia,
Wegener went on to suggest that at one stage these two were united in
a single land mass, Pangaea. The 20 per cent or so of the earth's
surface that is land stretched from pole to pole as one more or less
continuous continental slab.

Although the joining of Laurasia with Gondwanaland can be made quite tidily by bringing north-west Africa and North America together, there are a couple of regions where problems arise. The Caribbean is one: the Mediterranean is the other. Both are areas where there has been a great deal of upheaval, volcanic activity and earth movement in the last 70 million years or more.

Stretching eastwards from Gibraltar and across southern Europe into Asia is the Alpine-Himalayan mountain belt. The rocks involved within it seem to have been laid down in a sea of truly oceanic proportions, known as *Tethys*. That Tethys existed throughout most of the Mesozoic era and perhaps for much of the Palaeozoic too, seems beyond doubt. No sea of this kind was suggested for this region in Wegener's day and he neatly rounded off the eastern side of his Pangaea so that only a small oceanic embayment lay between Australia and India. Tethys appears on our maps of Pangaea today as a major eastward-facing gulf with India lying to the south of it.

How long was Pangaea in existence? Were the continents gathered into this cluster from previously scattered positions only to be split apart again? No one knows for sure, but the idea that the Alpine-Himalayan system resulted from the pinching of an open oceanic region implies that similar grand movements could have taken place earlier in the earth's history. Perhaps the (now denuded) mountain belts formed in mid and late Palaeozoic times between the Canadian and European Shields were raised by the collision of two continental blocks. There followed faulting on a large scale and volcanic activity. The situation then bears some resemblance to that in East Africa today where faulting and volcanoes are active in the rift valleys, a region where, according to some geologists, Africa is beginning to break up.

Maybe only for a short while, 50 million years or so, were Gondwanaland and Laurasia in contact. Pangaea may thus have existed only in Permo-Triassic times. How Gondwanaland and Laurasia themselves came into existence is even less certain. Gondwanaland seems to have been formed about 500 million years ago, the beginning of the Phanerozoic eon, and Laurasia perhaps since mid-Palaeozoic times. All kinds of guesses can be made about the mosaic of continental fragments in pre-Pangaea times.

During the last 60-or-so million years the break-up of Pangaea continued with continents drifting generally northwards and for the most part away from one another. Sea floor spreading was, and is, still, the mechanism underlying the drift. About half the ocean floor has been renewed in this era, which means that spreading has been at a geologically fairly rapid pace. In fact it means that virtually a third of the entire surface of the earth has been renewed from below via the oceanic ridges and the trenches. The shapes of the continents as we know them today begin to clarify and the great Alpine-Himalayan mountains rise from Tethys. In the Americas the Cordilleran ranges of the west coasts are pushed up and volcanoes rumble and cough out ash and lava on a truly enormous scale. For the first time New Zealand can be seen as a separate entity, broken off as Australia moves northwards.

Of course a greater part of the record of the Cainozoic era is available than of the earlier geological eras and it has perhaps claimed the most attention in respect of major structural features of the crust. Volcanic activity today is a legacy from the crinkling and distortion of the lithosphere and the generation of terrestrial heat that has been going on throughout the Cainozoic. The continents have assumed positions scattered over an area of the earth wider than any we can be sure of in the past. The effects, geological, geographical and biological, produce the great range of environments and provinces we have today.

North America continued to go relatively west in the Cainozoic and the North Atlantic rift has slowly spread into the Arctic Basin. Tremendous volumes of basalt have welled out at the surface while this split was widening and creeping northward. The Icelandic and Hebridean volcanic areas are, or have been involved in part of this process, and the ocean has acquired its distinctive shape. To the north-west of the Hebrides a large area of shallow sea floor, Rockall Bank, may have split from western Europe during this time, but, as far as we know, has never formed a truly continental unit on its own by emerging above sea level.

In the Caribbean area extensive movements were in progress throughout the Cainozoic, with widespread volcanoes in action. North and South America tended to rotate slightly relative to one another and threw up the arc-like lines of islands of the West Indies. With the help of powerful vulcanicity, the Isthmus of Panama was forged to connect the two continents. South America itself also continued to move westwards. Eventually it reached the position where it began to impinge upon the Andean Trench, displacing it, too, westwards. There must in the west have been volcanic and earthquake activity of quite horrific frequency and intensity during this encounter. The west coast of South America still suffers from these hazards, and may do so for a geological age or two to come.

Africa crept slowly northwards closing up the western end of Tethys against Europe by shoving up Tethyan sediments and chunks of underlying basement and raising the Alps as it did so. Arabia too pushed the Tethyan sediments up into huge ruckles against southern Asia. But a new feature makes itself apparent in this region during the Cainozoic. The great Rift Valley system of faults in East Africa, the Red Sea and Middle East areas, comes into being. Thus Arabia is split from the north-east corner of Africa proper and the great faults run through the African Shield southwards as far as Rhodesia. Minor rifts developed running north-eastwards across Libya to Italy and the Rhineland. They are truly great lines of cracks, paired, and producing the spectacular inward-facing scarps of the Rift Valleys. The narrow central portion between the faults has subsided during the Cainozoic and volcanoes have erupted over a long stretch of the rift area. Not all of the volcanoes are confined to the valleys themselves, some occur on the shoulders of ancient shield rocks round about. But their presence seems to indicate a local excess of terrestrial heat and the availability of other cracks and fissures by which they could erupt. The volcanoes of West Africa too must owe their origin to 'hot spots' developing below the shield rocks as movement took place within or below the great crustal plate on which Africa rests.

Perhaps the most spectacular achievement was the voyage of India ever northwards until it had excluded the last Tethyan waters from between its northern side and the southern margin of Asia. Squeezed into the narrowing space between the continents, the Tethyan sea floor sediments were piled into great stratal folds and bulges thousands of metres thick. The result has been the uplift of the mightiest range of mountains known today. The Tethyan Trench was now obliterated and only to the east of the Bay of Bengal do trenches swing around the south-eastern margin of Asia.

Even in this Indonesian region the trenches have not been free from squeezing. The northward push of Australia has ruckled them to the north and west of New Guinea. In its turn eastern Asia itself has tended to swing in a clockwise direction against the eastern-most parts of the old Tethyan trench system.

The continent of India in fact seems to have under-thrust the southern margin of Asia as a result of the opposing motions of the two continents. The positive isostatic anomaly of the southern Himalayas is produced by the double thickness of continental crust caused by this under-thrusting. At a fairly early stage in the Cainozoic the Indian continent seems to have encountered some feature in the mantle which released basaltic magma and spread the Deccan plateau basalts over so much of that continent.

As Australia fled to less southerly latitudes the rift between it and Antarctica was opening up and New Zealand splintered away from Australia. Antarctica itself tended to move but slightly and that was in a slow anticlockwise swing around the polar centre. This movement was not fast enough to keep pace with the westward drift of South America. In trailing behind the southern tip of the Andes and the horn of Antarctica during part of this movement the great loop of the Scotia Arc was formed. Antarctica had become separated from the other continents and isolated as a sixth plate in the crust by the development of the ocean ridges sometimes known as the Pan-antarctic Rift System. Alfred Wegener recognised the westward drift of the continents and believed that they had also moved away from the poles. Evidence today certainly confirms this first conclusion but seems to suggest that, except for Antarctica, the continents have fled northwards.

A-roving We Will Go

Much of the early evidence for continental drift was based on the distribution of fossils. By reassembling the portions of Gondwanaland and Laurasia many of the anomalies seemed to be removed. Wegener felt at the outset of his work on drift that land bridges were not satisfactory in explaining some of the peculiar distribution of fossils. Having now accepted the theory of continental drift as respectable, if not a proven fact, we might reflect that it can possibly explain the odd distribution of some groups of plants and animals in today's world. We might view the continents as super-arks floating about for millions of years with evolving animal and plant cargoes or passengers. At one stage the passengers are gathered into one ark, and they—or some of them—troop off to another that hauls alongside. Separation of the arks means that some forms that have migrated from one continent to another may have quite different and new conditions to contend with from those they had previously. The consequences could be favourable or they could be disastrous. Let us look at a few examples.

Pine trees belong to the group of trees known as conifers. It has recently been discovered that from late Cretaceous to early Tertiary time different types of conifers lived either in Gondwanaland or in Laurasia. No single species lived in both continents. Conifers do not seem to survive dispersal across oceans but during the Cainozoic the conifers from Gondwanaland spread into the northern land masses. For example, a southern form known as *Podocarpus* has spread into Central America, Indonesia, Asia and Japan from Gondwanaland. The northward movement of India to join the southern margin of Asia could have been vital to the wider dispersal of conifers to the north-east.

There are other instances of the strange distribution of certain plants. One group is found in Madagascar and again in north-eastern South America, but is not to be seen anywhere in the African continent in between. Here it is suggested that when Gondwanaland was intact the group was spread across the entire area but following the break-up of this super-continent the plants became restricted to progressively smaller and more widely separated areas until they died out in Africa and remained only near the eastern and western limits of their previous domain.

Perhaps the most intriguing story of continental drift and the dispersal of animals concerns the pouched or marsupial mammals. These beasts are represented today by kangaroos, koalas and other animals in Australia and by the quite distinct opossums and some others in the Americas. No other continents have marsupials today.

Fossil marsupials are, however, found over a wider area and they appear to have originated in North America in the Jurassic period. The placental mammals appeared on the scene at some time between the late Jurassic and the mid Cretaceous periods and by the end of the Cretaceous the marsupials were in retreat before the placental mammals. Many of them withdrew into South America and placentals went there too. One wave of mammals entered the southern continent at the end of the Mesozoic, a second followed in early Oligocene and a third since the Miocene. Each one has displaced the previous inhabitants.

The Australian marsupials have clearly not been affected by such waves of new invaders. Marsupial evolution in Australia has been virtually uninterrupted throughout the last 70 million or so years. To reach this however the marsupials began their explorations out of the Americas into Antarctica and on to Australia between late Cretaceous and early Cainozoic times. The business of getting from one continent to another may have been rather hazardous because open water may have existed from time to time between them. By 'island hopping' where no continuous land connection existed the marsupials may have been able to overcome this hindrance. Having got to Australia, alone or in company with placental mammals, the marsupials spread into all the available ecological niches from Oligocene time onward. No placental mammal fossils are found in Australian rocks so we cannot be sure that those mammals ever reached the continent. But the mammalian 'explosion' elsewhere suggests that the placental mammals were so successful that an absence of fossils means that they did not put in an appearance in Australia.

Occupying such a far southerly position as it did, Antarctica probably had 'polar seasons' in that there was a long period when the hours of daylight were few and a long period when they were many. We have no reason to believe that it was really cold until late in the Cainozoic.

Wegener did not live to see the acceptance of his hypothesis by the entire geological fraternity. The few adventurous minds that advocated the recognition of continental drift, despite criticisms, were anxious to show that there was a means by which the great internal forces of the earth could be adapted into the task of shunting continental crust from place to place. Even now not every geologist is convinced that continental drift has taken place. Several lengthy accounts attack the validity of the evidence upon which the theory is founded—some of them without even mentioning the existence of Wegener's work. While it is true that much of the evidence may be interpreted in other ways, the great bulk of it seems to demand that drift of a kind has taken place over the last 200 million years.

Pl. 17.   A significant crack in the crust is the San Andreas fault, separating a
northward-moving wedge of California from the rest of North America. This photograph
of part of the fault in southern California shows that the hilly area to the left (south) of
the fault line is moving upward (westward) relative to the plain on the right. The offsets of
the two large water courses as they cross onto the plain are clearly visible.   *Courtesy of*
The New York Times

Pl. 18. The passing of an ancient glacier is marked by the characteristic ice-scoured and grooved rock surface is the Hoggar region of the Sahara. Discovered by French geologists within recent years, this kind of evidence reveals that an ice sheet of continental proportions existed in this area in late Ordovician times. *Photograph from* Les Grès du Paléozoique Interieurau Sahara *by S. Beuf, B. Biju-Duval, O de Charpal, P. Rognon, O. Gariel and A. Bennacef, Courtesy of Éditions Technip*

Pl. 19. Coalseam Cliffs in the Theron Mountains of Antarctica are composed of coal-bearing rocks of Permo-Carboniferous age and are intruded by massive sills of slightly younger dolerite, a basaltic igneous rock. The coal rocks and the dolerites can be closely compared with Permo-Carboniferous rocks in South Africa and they provide much relatively new and convincing evidence of the previous closeness of these two continents. *Courtesy of British Antarctic Survey*

Pl. 20.   Fossils of creatures with shells or bodies like those alive today are easy to interpret but this little animal, *marrella*, is another matter altogether. Hundreds of these tiny sea creatures were fossilised in a fine black mud in Cambrian times perhaps 550 million years ago.   *Courtesy of H. B. Whittington*

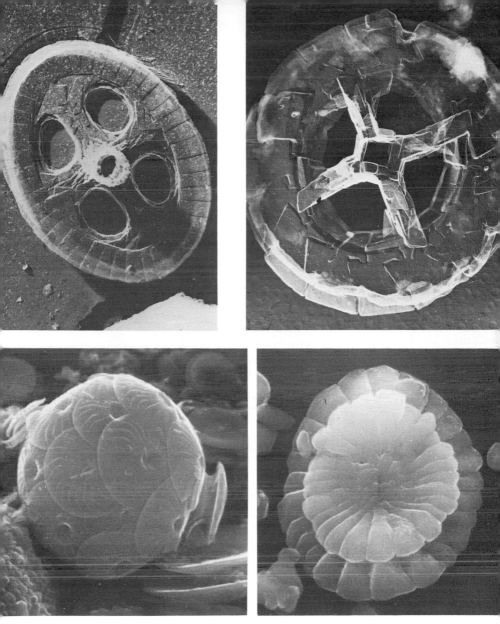

Pl. 21.   Known as coccoliths, these tiny fossils are composed of calcium carbonate and in countless millions make the pure whitish limestones we call chalk. Late Mesozoic and early Cainozoic chalks are extremely widespread and the tiny marine algae which produced coccoliths must have been many times more abundant than they are today. The circular specimen at top is ×20,000 and the oval ×12,000, Below right is a modern coccolith (×14,000) and at left is a modern coccosphere composed of many overlapping coccolith plates, from a deep Atlantic ocean floor sediment (×6,700).   *Courtesy of M. Black (above) and B. M. Funnell (below)*

Pl. 22. *Above* This narrow valley in Lebanon possibly marks the line of contact between the African and the Arabian plates (fig. 1–5). The African plate underlies the distant hills: the Arabian plate is in the foreground. Very possibly movement is still going on between these plates, just as it has for much of the Cainozoic era, obliterating the Tethyan Ocean. *Courtesy of P. L. Hancock*

Pl. 23. *Opposite* Excavating soil on the moon, this Apollo 15 astronaut is seeking samples and information about the nature and origin of the soft thin film of unconsolidated material covering the rock beneath. At a depth of 30 to 35 centimetres the trench had to be abandoned because of a hard layer that could not be penetrated by the scoop. Soil-forming processes on the moon are quite different from most of those operating on earth but the rocks themselves are not dissimilar from many basaltic rocks on earth. *Photograph from N.A.S.A., Apollo 15 programme*

Pl. 24.   Space travel will undoubtedly reveal many new clues about the origin and history of the solar system. Despite all the new remote control systems for space vehicles, the most exciting exploration in space is still that performed by men on their own two feet or on a 'moon buggy'. Exploration of other planets by anything like these means is still a long way in the future.   *Photograph from N.A.S.A., Apollo 15 programme*

CHAPTER IX

A Hidden Mechanism

WITH such a weight of geological evidence behind it the theory of continental drift might have been regarded as secure from criticism. Yet formidable problems remained to be solved and geologists in the period 1930 to 1960 were largely without the means to do this. Most important were the questions of what had provided the motive power behind the movement and how it was that the leading edges of the moving continents were not more deformed and buckled as they overcame the resistance of the oceanic crust. A frequently quoted analogy was that of scum floating on simmering jam, and the continents were indeed thought of as floating blobs moving as a convecting layer beneath them rose, spread and sank.

Many geologists claimed that they had evidence of lateral movements in the structure of the continents and believed that the great episodes of earth movements in the histories of the continents could be explained in terms of the effects of convecting currents beneath the crust. One of them was Arthur Holmes in Britain. His belief in the efficacy of convecting currents in the mantle as a means of causing geosynclinal subsidence, orogenic compression and subsequent isostatic uplift was firm and his view of the cyclic nature of the currents and their surface effects was clear and precise. This was well before anything was known about the activity of the oceanic ridges and the configuration of the ocean floors.

At about the time that the new data on the oceans was being published S. K. Runcorn, Professor of Physics at Newcastle-upon-Tyne, came forward with an idea that continental drift, convection in the mantle and the growth of the earth's iron core were connected. Recalling that in the first 500–1,000 million years of the earth's existence the fluid core was forming from the separation of iron and its inward migration, he suggested that at first there was within the mantle a single large convecting current. While this stirred the mantle for millions of years the crust formed as a single huge continent. As the core grew in size from the rain of iron from the outer layer the mantle became correspondingly less thick. It could no longer maintain a single powerful current, and several lesser separate currents replaced it. The large continent was split and the pieces continued to grow, the convecting cells became shallower still and more numerous. Continental drift would be caused by the change in these patterns of convective currents in relatively recent geological time.

It is a clever idea which implies that crustal movement has always been an effect of convective activity in the deeper layers of the earth. Later information suggests, however, that once the core had taken shape at an early date in earth's existence its size increased but very little indeed. Nevertheless, Runcorn felt, as Holmes had, that convective cells within the earth must exist and must have a long-continuing effect upon the shapes, positions and deformation of the continents.

Several continents may indeed have suffered deformation, especially at their edges, but viewed overall they have generally behaved as rigid slabs or rafts rather than soft patches of granitic scum on a basaltic jam. A currently more popular and accurate analogy of the crust of the earth is one of ice rafts or plates floating on water. The entire surface of the planet is now thought of as comprised of six or possibly seven major rafts or plates of which the continents form only part. Between some of these major plates there may be as many as twelve much smaller platelets. The plates are in motion, carrying the continents with them one way and another, and some parts of the plates are being destroyed while other parts are being newly created. It is the growth, movement and destruction of these plates that has controlled continental drift and orogeny—or at least so says the new theory of 'plate tectonics'.

This theory is a remarkably useful one and in its simple outlines presents us with a nice means of better understanding the behaviour of the crust. The story of how the theory evolved is remarkable in the degree to which it involves geologists, geophysicists, hydrographers and scientists of other disciplines.

As we mentioned in an early chapter, the latter half of this century has seen a great surge of interest in the study of the sea and the sea floor with a battery of new instruments and new techniques. The Americans, with their renowned oceanographic institutes at Woods Hole, Massachusetts, and La Jolla, California, were foremost in this work. The Woods Hole teams, led by Professor Maurice Ewing, discovered that the crust beneath the oceans is a mere 6 or 7 kilometres thick in contrast to 30 to 40 kilometres beneath the continents. This was done by finding out how the shock waves from large explosions (such as depth charges) travelled through the submarine crust.

They also discovered that the mid-oceanic ridge found in the North Atlantic extends into a world system of such ridges, pervading all the oceans. You will recall that the ridges are sharp rises of 2 or 3 kilometres above the abyssal plains, and are scarred by a median rift valley fault system. Volcanoes and earthquakes are characteristic of the ridges, and are good indications that the ridges are unstable

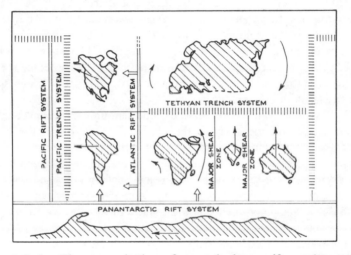

Figuré 9–1   The general plan of crustal plates, rifts and trenches today. The direction in which the rifts are migrating are shown by open arrows. Black arrows indicate the directions in which the individual plates have moved. (After Dietz and Holden.)

places, marking zones where the earth's internal energy is constantly breaking out.

Recognition of these phenomena in the oceans of the world led Professor Harry Hess to propose his theory of 'sea-floor spreading', to relate them to a variation of the older convecting current idea. He saw the ridges as being sited over the upswelling parts of convecting currents in the mantle. The thin ocean crust was produced by a physical and chemical alteration of the mantle as it is pushed outwards from the crest of the ridges. In short, the sea floor spreads, is continually being pushed outwards from the ridges by the upswelling of the mantle from below, causing intrusions and volcanoes more or less incessantly. This could push the continents apart and on the basis of our knowledge of North Atlantic geology, Hess thought a movement of 1 centimetre per year outwards on each flank would not be far from right. We now know that about 2·9 cubic kilometres of new ocean floor are produced at the ridges each year around the world and that this rate has not changed much since Cretaceous time.

But if the sea floors are spreading in some directions, either they must also be consumed in some regions or else the earth is expanding. Moreover, to complete the convecting cell there has to be a zone where material has cooled enough to descend back to the lower levels once again. If the mantle came up at the ridges where might the same amount of material be returned to the depths? The answer seemed fairly obvious—the great oceanic trenches, as in the Pacific, are the locations of the descending currents where ocean crust is drawn down and resorbed into the mantle.

This simple framework proved good enough to support other observations and ideas about the chemical evolution of the ocean crust from the mantle below, and its return to the mantle. Such sediment as collected on the oceanic crust would be gathered slowly into the regions of the trenches but being light and plastic enough to resist being drawn below would form island arcs or new lands welded on to the adjacent continents. Another facet of the idea was that if this activity has been going on since the crust first formed, most of, if not all, the water of the oceans may have been derived from the volcanic breath of the mantle.

Whereas geologists in the past had thought that the violent and significant events in earth history had all primarily been associated with the continents, Hess was suggesting that it is the oceanic areas where most of the action has been. Continents have been pushed around at the behest of the oceanic parts of the earth which are constantly changing. To support this there was the remarkable fact that no deep ocean floor or (volcanic) oceanic islands yielded rock more than about 100 million years old. The present ocean floors are really rather young geological features. The 1960s were to bring enough new evidence to confirm all this and more in the same vein.

A Magnetic Tape Recorder
Basalts, as we have seen, are weakly magnetic rocks and the ocean floor has a superabundance of basalts. It was natural then that the early surveys of oceanic ridges included searches for anomalies in the

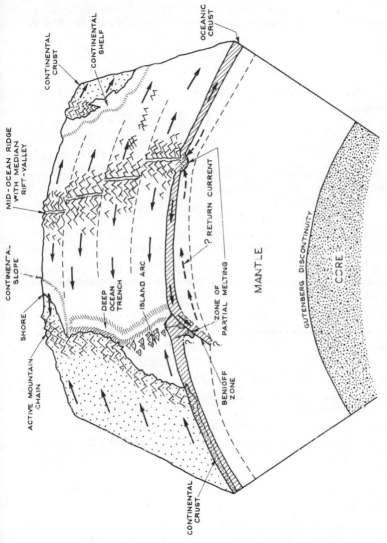

Figure 9-2   The main points in the concept of plate tectonics are shown in this diagram of a hypothetical segment of the earth. The arrows show the motion of the lithosphere plates relative to the mid-ocean ridge. (After R. Mason.)

earth's magnetic field associated with these features. Sure enough, some surprising results and correlations were to be achieved. In the North Atlantic there was found an appreciable increase in the strength of the earth's magnetic field over the median rift valley along the ridge. And the increase persisted even where the valley itself faded away—it was a strong, linear magnetic feature.

There were others to come. From the north-east Pacific came a record of more linear disturbances in the earth's field. They run north–south in strips several kilometres wide and in at least one zone the pattern is offset by an east–west fault or submarine fracture. It is possible that the offset represents a lateral shift along such a fault, but nothing quite like this pattern has been found on any continent. Along one such fault, the Murray Fracture Zone, the shift is as much as 150 kilometres.

Studying submarine magnetic anomalies such as these, two young researchers, F. J. Vine and D. H. Matthews, came up with the idea that the linear magnetic properties of the ocean floor are acquired when the basalts are first crystallised from magma. At this point we should recall that the phenomenon of 'reversed' as well as 'normal' polarity can be found in rocks that have 'fossilised' the earth's magnetic field at the time of their formation. Vine and Matthews suggested that when basalt is added to the sea floor at the oceanic ridges it acquires a magnetic 'print' of the earth's field. When, as happens from time to time, the earth's polarity is reversed this, too, is 'recorded' when new basalt crystallises. As time passes, the older basalt is pushed laterally away from the ridges by the new, and the ocean floor bears the imprint of the magnetic field that occurs, normal or reversed, over millions of years. Since new sea floor is, so to speak, added in strips at the ridges, the magnetic points, also, are strip-like, linear phenomena. And the points are roughly paired, each one having a near twin on the opposite flank of the ridge, reflecting the spread of new floor away from the ridge on each flank. Each flank has its magnetism 'taped'.

A few years' work in the mid 1960s brought striking confirmation of this idea. Recent basalt flows from the ocean floors were studied and their ages computed. From these a time scale or calendar of magnetic reversals during this period was indicated. To the immense satisfaction of all concerned, there was confirmation of this calendar from a nearby source. The thin skin of sediments formed on the ocean floors was also sampled and tested for its remanent magnetism, and the work produced an identical calendar of reversals for the last few million years.

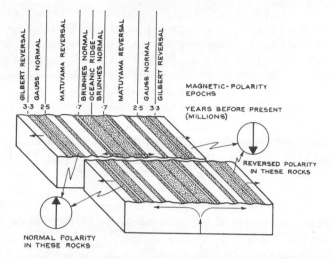

Figure 9-3   A magnetic tape record of sea floor spreading. On each side of the mid-oceanic ridges are corresponding strips of basalt floor in which the magnetic particles are aligned in the way they do when basalt crystallises today (dots) and strips where the particles align in the reverse direction (no dots). The diagram shows these strips and the names given to the spans of time during which they were formed. Such splendid (though here diagrammatic) symmetry suggests that the rocks welled up in a more of less molten state and gradually moved outward – sea floor spreading. The displacement of the two blocks represents a transcurrent fault or fracture zone. (After R. S. Dietz.)

As more and more sea floor rock and sediments have been examined, it has been seen that the rates of spreading have varied from place to place. For example the spread from the ridge in the North Atlantic is about 1 centimetre per year near Iceland, but is as much as 9 centimetres per year on the flank of the East Pacific Rise off Ecuador. Most spreading seems to be 3 centimetres or less per year. Patterns of magnetic anomaly on one ridge flank can be seen to correspond roughly but convincingly with those on the other in each oceanic ridge, and there is an almost world-wide correlation between the ridges.

An interesting—and expensive—exercise was devised to put these ideas about palaeomagnetism and sea floor spreading to the test. By assuming constant rates of spreading about a ridge, one can predict a date for the ocean floor at any given distance from the ridge. Sediment samples and rock grabbed from the floor could provide confirmation but rock core drilled from the living basalt would give much more reliable data. With characteristic speed and thoroughness American scientists were to mount their Joint Oceanographic Institutions Deep Earth Sampling project (JOIDES), an ambitious plan to drill holes at various points along traverses across the oceans. We mentioned in Chapter I that the work is being carried out from a remarkable ship, the *Glomar Challenger,* and rocks of all ages back to middle Jurassic have been encountered. In terms of scientific return, this expensive and difficult deep sea operation has been tremendously successful. It has confirmed that ocean floor spreading seems to have gone on throughout this period of 165 or so million years, and that during the Cainozoic the mid-Atlantic has undergone spreading at about 2 centimetres per year on each side of the ridge. Adding all the areas of ocean floor produced on the flanks of the world ridge system during Cainozoic time, which is after all only 67 to 70 million years, it appears that almost half of the ocean floors are involved. In other terms, about one-third of the earth's surface is less than 70 million years old and has been created in only the last 1·5 per cent or so of geological time.

Plate Tectonics

These are quite spectacular concepts and have been formed remarkably quickly in comparison with the rate of progress in understanding the earth in the previous century or so. But the 1960s were to see a further development of ideas on the nature and history of the crust. This has produced the concept known as 'plate tectonics', and it has had a most profound effect upon geological thinking.

Several geophysicists were impressed with the great transverse fractures that became known when the oceanic ridges and floors were explored. J. Tuzo Wilson, doyen of the Canadian geophysicists, and an early advocate of ocean floor spreading, named them transform faults, as they cross the ridges approximately at right angles apparently displacing one side relative to the other by as much as hundreds of kilometres. He thought that the offset of the ridges might be original features that have not changed as spreading has gone on. Seismic activity along these lines appears to be confined to the region between the end of the ridge broken by the transform fault.

Meanwhile records had been accumulating that showed that earthquakes originate at great depth beneath the mountain chains and ocean trenches thought by Hess to be the 'sinks' for ocean crust. The foci of earthquakes below the ocean ridge crests, on the other hand, are shallow, less than 20 kilometres down for the most part. Elsewhere, there is hardly any seismic activity at all. The earthquakes must be caused by friction and the friction is confined to the narrow zones Hess picked out. To several geologists confronted with these facts they suggested that the crust of the earth was analogous to a set of rather rigid plates, the boundaries of which are formed by the oceanic ridges, transform faults and young mountain chains. At the ocean ridges new crust is added to the plates, at the trenches and beneath the mountain chains plates are in some way destroyed or drawn back into the lower regions. The evidence from the earthquakes seemed to confirm this beneath the trenches and mountains. And all the plates must be in motion or the ridges are moving relative to the centres of the plates.

The earthquakes beneath the trenches and adjacent mountains seem to fall along a zone which descends beneath the edge of the continent at an angle of about 45 degrees. Could this be the zone down which the oldest part of a plate moves as new plate is produced at an oceanic ridge? It is a welcome suggestion and it lets us off the hook of having to propose that the earth has been continually expanding from a state when the continental crust covered the entire earth and when there were in fact no ocean basins at all.

On the map of the world, or the globe, then, we can plot the boundaries of the six or seven great earth-covering plates and rather more smaller plates. They are thought of as being relatively thin—no more than 150 kilometres thick—and internally rigid. They are not all the same size and they are all in motion relative to one another. It is at the boundaries of these plates that virtually all of the volcanicity, earthquake activity and mountain building has been, and is, going on. As you can see on the map, the emphasis has shifted from the edges of the continents to the edges of the plates as being the most important regions in which to observe how the basic processes moulding the configuration of oceans and continents take place. Of course there are areas, such as parts of the western margins of the American continents, where continental and plate boundaries largely coincide. In the geological past other continental margins may have found themselves in similar situations.

In no time at all the proponents of the theory of plate tectonics presented the public with the inescapable observation that there are three types of plate margin. One is where new crust is created at oceanic ridges; it is known as a constructive margin. Another type is destructive, as in the oceanic trenches, in that two plates are approaching one another and one is pushed, or slips, down beneath the edge of the other at an angle of about 45 degrees. In the third type of margin the movement is lateral as one plate slips and grinds its way past another. This is the case, we think, along the Californian coastline where the American plate slips by the Pacific plate.

Plates and Mountains

So our ideas about the behaviour of the crust of the earth are now very different from the old concept of a contracting earth with wrinkles in its skin like a drying apple. Arthur Holmes, Alfred Wegener and many other imaginative but cautious geologists had held that whatever was the driving power behind continental drift was also the power behind orogeny. Perhaps orogeny only took place when drift movements were going on, or perhaps drift was a result of orogeny.

The plate tectonics theory helps to explain the movements of the crust. It gives us an explanation not only of drift but also of how certain kinds of mountains and continental margins have come into being over the last 200 million years or so. Sticking to our uniformi-tarian guns, we ought to be able to use the idea to explain older geological features and events. The evolution of the ancient fold belts might be revealed by applying the idea to the study of their remains. In due course it might even be possible to unravel some of the mysteries of Precambrian continent evolution by these means. The prospect is positively exciting, the horizon unlimited.

As we watch the plates today, they are all in motion relative to one another, sliding past, under or over the margins of their neighbours. At the oceanic ridges the plates grow and spread away from the ridge, at the oceanic deeps they are consumed. The great mountain ranges of the present lie adjacent to these destructive margins. The Circum-Pacific belt occurs where oceanic crust meets continental crust. The Alpine-Himalayan belt lies where two plates with continental crust have collided. So perhaps the explanation of mobile belts is that they develop at the destructive margins of plates.

Where continental crust occurs it lies above oceanic crust, floating on it and resisting any downward drag that moving oceanic crust might exert. Here the analogy of ice on a slowly moving mass of water comes to mind again. Thus oceanic plate moves under the edge of continental plate, but of course not without plenty of side effects being generated. As the oceanic plate moves beneath the edge of the continental, the accumulated sediments on the plate margins are squeezed and metamorphosed, as we shall see shortly.

Where two masses of continental crust are brought together the side effects are nonetheless severe, but they are rather different. Here we can conjure up again the idea of a great vice, the jaws of which are the' segments of continental crust, closing slowly and squeezing the soft, light sediments that accumulated on their edges and on the oceanic floor as they moved.

The meeting of a plate with continental crust at its edge and one without is a major event in earth history. How many times it has happened is not known, and of course it can only happen when the continental mass is at the leading edge of a plate. Today the western seaboards of the American continents reveal what happens. The ocean plate is warped and ground beneath the buoyant edge of the continent and the sediments on top of the ocean plate are scraped into a pile like wood shavings before a plane. The dipping contact zone between the two plates is, in fact, the Benioff zone. The heat of the interior and from the friction between the plates melts the local rock and, magma forms from the local crust and results in volcanoes spitting out andesitic lavas and ashes. As the production of heat continues it may reach a point where some of the sialic continental plate itself also becomes molten. This molten material tries to escape upwards and granitic intrusions are produced. So along the narrow zone at the edge of the continental plate fierce heat arises and with it all means of igneous and metamorphic activity, fracturing and folding, take place. We have an orogen, and it may last for many million years.

As our continental plate butts against the oceanic plate the margin may thicken and warp. The region is uplifted, vigorous erosion follows and new suites of coarse sediments are produced. Many of these pour into the trench on the ocean side of the margin and may in turn become incorporated in the next orogeny.

Much of this account suggests a process that could produce the features essential to the geosynclinal concept. The growth of a thick wad of sedimentary material and its subsequent fate in the moils of an orogeny might be due to an oceanic floor carrying sediment to a destructive plate edge to be scraped into a contorted buoyant mass as the floor descends beneath the adjacent plate. Geosynclines, or at least some of them, can be interpreted as occurring where plate destruction is in progress, and their contents may in part have been the covering of an ocean floor rather than the infilling of a single relatively narrow long-lived downfold. Geologists are not inclined to abandon the geosynclinal idea solely because of this, but they do see that it demands a major reconsideration of the history of many a geosyncline. It is perhaps appropriate that a concept which first emerged as a result of the study of Appalachian geology should be modified so much by the work of geologists from the north-eastern U.S.A. several generations later.

Continental Collisions

The effects of continental plate margin and oceanic plate margin coming into collision are spectacular enough, and the results when continent meets continent are scarcely less so. In these instances the actual 'bump' is preceded by a long phase when the leading edges approach one another. As they get nearer the sediments that have collected on the ocean floors may be scooped and scraped together; the strata of the continental shelves become pushed together and the jaws of the vice begin to close in earnest. From here on orogeny takes place as elsewhere. But while there may have been a trench or oceanic deep between the two continental edges for at least part of the time the plates were drawing close, the open ocean floor may also be involved over much of this time.

When Tethys existed between Laurasia and Gondwanaland it probably retained oceanic proportions for a time. Then, as the southern super-continent split and the fragments drew north upon the flank of Eurasia in Mesozoic and Cainozoic times, the sea floor diminished, Tethys shrank. Eventually the Alpine-Himalayan belt was squeezed up from the sediments of this lost ocean. Bits of oceanic mantle may have been pinched into the orogen to rise like great basic blood blisters. Chunks of the continental basement were sheared off and pushed their way up through an overlying cover of later sediments. At first sight the cross-sections drawn by the early Alpine and Himalayan geologists seem rather chaotic, but they clearly show structures that have grown in response to strong pressures from the south. The tectonics of this belt are indeed as complicated as any in the world, but this is what one might expect when such major crustal movements as the wanderings of continents are abruptly halted by collisions. As investigations continue in these rather demanding terrains, more and more sense is being made of the structures discovered there.

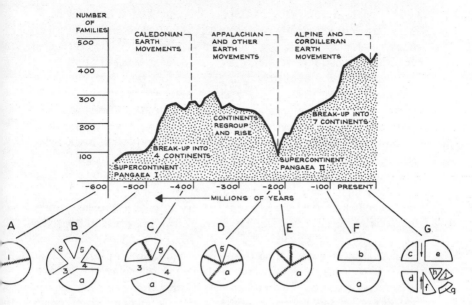

Figure 9-4 The diversity of living things at any time in the history of life over the last 600 million years seems to reflect the pattern of continental assembly and break-up. When the continents are clustered together diversity is low: when they drift apart the diversity is high. Diversity is indicated on the graph by the number of different families or organisms. 1, Pangaea 1; 2, Pre-Caledonian Ocean; 3, Pre-Appalachian Ocean; 4, Pre-Hercynian Ocean; 5, Pre-Uralian Ocean; a, Gondwanaland; b, Laurasia; c, North America; d, South America; e, Eurasia; f, Africa; g, Antarctica; h, India; j, Australia. (After Valentine and Moores.)

It is not improbable that such collisions took place from time to time in Precambrian eras. We have detailed knowledge only of one major Cainozoic collision. The possibility that some of the big orogenies of the Palaeozoic, say the Caledonian–Acadian orogenies of eastern North America and western Europe, were produced in this way is very real. We return to it in the next chapter.

There must, however, have been a time before plate tectonics along the lines we now recognise came into effect, or when the system of movements within the asthenosphere had different effects from those taking place in the last few hundred million years. Rather far back in the Cryptozoic eon the level of heat and energy within the earth would have been high enough to maintain movement at a faster pace. A more turbulent expression of this energy might have taken place with many small plates being involved, and very probably the crust itself was hotter and thinner; when one plate pushed beneath another the friction may have completely disorganised the upper plate. No simple long and narrow belt of crumpling, metamorphism and igneous irritation would result but a much more widespread and wholesale heating up of the undersurface of the top plate may have occurred.

The sutures or scars formed where ancient continent was welded to continent as a result of crustal drifting are not easily recognised and small ancient sutures may well elude our attention. It is quite possible that the results of such a very ancient form of plate tectonics would have been obliterated by the metamorphism and orogeny of later times. Previous to about 3,000 million years before the present, relatively close proximity of the moon may have had severe effects upon earth's crustal formation, disrupting or modifying the flow of plastic or fluid material in the asthenosphere and rupturing the lithosphere.

To what extent the tidal influence of the moon has ever affected the asthenosphere and its lithospheric plates is unknown, but theoretically it could be appreciable. Presumably it would pull the continents slightly westwards in the course of time. The rate may have been uneven because of the differences in their sizes and thickness of the lithosphere. The influence may be strong enough to promote stresses sufficiently within the lithosphere to produce the strong north–south aligned rift systems of the oceanic ridges, Africa and several remarkable rifts and faults in the Americas. It has also been claimed that the tendency of the plates in existence today to move round in a clockwise direction is due to a lunar tidal influence.

Summing Up

The idea of crustal plates moving, evolving, submerging is all so relatively new and revolutionary that it may help if its main points are repeated before we see how it may impinge on another part of our story, the evolution of life on earth.

The mid-ocean ridges with their rifts and great volcanic and seismic activity mark the belts along which underlying mantle currents rise. They can migrate over long periods of time, extending and changing their shape as the Atlantic ridge has. On the other hand, the deep trenches, marking the regions under which old crust is resorbed into the mantle, probably tend to remain in fixed positions. Nevertheless, when the leading edge of a drifting continent came up to a trench it may have displaced it. The Andean trench seems to have been pushed westwards as the bulge of the western coast of South America came up against it from the east. Most of the great deeps and trenches are arc-like in plan and their migration always seems to be in the direction of the convexity. In the case of the Japan Sea the trench is seemingly moving away from Asia.

As the plates grow, push and jostle each other they can change in shape and even be entirely destroyed by resorption. This will happen where an oceanic rift-ridge system migrates into a trench. One example of this might be where the north-east Pacific plate has been almost entirely resorbed. During the shoving and sliding that the plates endure, the rifts, trenches and the great transform faults may all be considerably modified.

When a plate is resorbed it is oceanic crust, sima, that goes under. The continents are buoyant and stay up on top. Sialic crust never seems to disappear in the way that oceanic crust does. New crust is always sima, never sialic continental material.

There seems to be nothing to suggest convincingly that the total surface of the earth is expanding so all new areas of crust are matched by the resorption of an equal amount of old crust at a trench. It is a case of what comes up, must go down. The movement up, the lateral motion and the sinking of old plate result from convecting movements in the asthenosphere below.

In 1971 two American institutions, the University of Hawaii and Oregon State University, were granted funds by the U.S. government, to examine the fate of a crustal plate from its birth in the East Pacific Rise in the Pacific Ocean to its destruction beneath the deep ocean trench off Peru and Chile. Scientists from Latin American countries also will be taking part in this major geological enterprise, involving marine geological dredging, grabbing and core-drilling, and many geophysical traverses to find out as much as possible about the Nasca Plate. Their work may provide information about earthquakes originating near the trench, and about the possible use of the trench as a waste disposal area. The shades of the members of the cruise of H.M.S *Beagle* are likely to watch these proceedings with a certain interest; it is 150 years since they fished in these same waters.

All in all, the concept of plate tectonics seems to explain the origin and nature of the major features of the face of the earth better than any other hypothesis has so far. Rifts, oceanic ridges and trenches, island arcs, folded mountain belts, volcanicity are all explained and their roles in earth's behaviour recognised. Nevertheless, the foundations of the concept are perhaps slender and need much more study—the 'fit' of the continents, the magnetic characteristics of the ocean floors about the ridges and the palaeomagnetic data from the continents are all rather poorly known. It is quite impossible to say with certainty how far back into time we shall be able to find evidence for lithospheric plates. Plate tectonics as described here, though, has an appeal in helping us to explain so much of earth's activity; and that is not to deny that there are other processes affecting the evolution of the crust. In the course of time we shall learn to what extent the idea really is as good as it appears now.

Plates and Populations

In the few years since the idea was first published nearly every kind of earth-scientist has been interested in the plate tectonics band-wagon and palaeontologists have been among them. They seek in the theory an explanation for the great spasmodic or episodic increases in the numbers of different kinds of organisms, and perhaps for the sudden widespread extinctions too. In the growth, movement and splitting of continents may lie the key to several problems concerning the history of life on earth. The movement of plates and the drift of continents means that the shapes and positions of the ocean brims, continental shelves and the shorelines of the continents must also have been changing, sometimes quite rapidly. During the last 600 million years or so this could have had a considerable effect upon the habitats of marine animals and the level of the sea itself. The Americans J. W. Valentine and E. M. Moores have offered the opinion that the diversity of marine life, or at least animal life, and fluctuations in sea level can be related to the breaking up and reassembly of continents.

In the super-continent Pangaea they saw a situation favouring the development of an extreme climate. It had very hot summers and very cold winters and the seas round about also had great seasonal changes as the interior lands warmed and cooled. In winter and spring the winds would have been off-shore. This condition today causes the cold lower waters off the continental shelf to rise as the warmer surface waters are blown away from the land. The cold waters are rich in the dissolved mineral salts which are essential to marine plant life; as a result there is often a 'bloom' of marine organisms at this time of year. In summer the whole process tends to be reversed with the sea winds drawn into the hot continental interior. The same state of affairs may have taken place around Pangaea. A kind of super-monsoon climate would then have prevailed.

At the same time the kinds of coastline around Pangaea would tend to allow a fairly unhindered migration of marine creatures from place to place, especially if the climate was much the same over wide stretches of the coast. The Mesozoic faunas do rather show that this freedom to migrate existed.

Now when the super-continent broke up each daughter continent would tend to acquire a marine climate of its own. The differences between those of sister continents might become fairly conspicuous, as they are indeed today. One might reasonably imagine that the climates of the smaller land masses would be more equable. The seasons would not be so sharply marked and extreme. In time the changes in the coastal waters would not be so large. The continental shelf seas would be chemically more stable, less prone to fluctuations in their content of nutrient salts with the seasons. We know that in modern shallow seas where these conditions are stable there is a larger diversity of species than in waters which range from warm to cold each year.

So would it have been in the past and around the daughter continents once they had separated. We see in the fossil record that there has been a tendency for each continent to acquire its own distinctive coastal faunas as time has gone on. The result of course is an increase in the total number of species as the continents have continued their separate existences.

Where smaller continents are brought back into contact with one another to make a larger land mass again the whole process could be reversed. It could be reversed to the point where some species would no longer be able to compete for survival. A reduction in the diversity of the plants and animals present would occur. Those kinds of organisms that were least adaptable or preadapted to fluctuating conditions would perish more than others. The very specialised forms would soon decline in numbers. Add to these factors those of climate dependant upon latitude, and a complicated but not indecipherable situation results.

We now begin to see that during Phanerozoic times there have been periods when super-continents broke up and periods when the fragments were regathered into one or two super-continents again. Faunal diversity, said Valentine and Moores, is related to these events. The Phanerozoic eon began perhaps with a single continent, Pangaea, in process of breaking up. By the end of the Cambrian period, it had split into four. The intervening seaways became the sites where the sediments of the Appalachian, Hercynian and Uralian geosynclines were to form. In effect, of course, these seaways became true oceans; no doubt they were wide and floored with true basaltic ocean floor.

The Appalachian ocean began closing from Ordovician times onwards and by the early Devonian it had been completely squeezed out of existence in the north. The Caledonian orogeny represents the final phases of this squeezing in Scandinavia, Spitsbergen, east Greenland, the British Isles and North America north of, say, New York. The southern part of the Appalachian ocean and the Hercynian ocean were closed in the late Carboniferous and Permian periods. The Uralian ocean had also disappeared by the end of the Permian period. Pangaea had emerged: the continents were reassembled once more into a single enormous land mass.

Pangaea, however, was short-lived. With the extension of the great ocean, Tethys, it split into Laurasia and Gondwanaland. Then in Jurassic and Cretaceous times the Atlantic Ocean made its appearance while Gondwana broke up further. From then on the continents have sidled off into the positions they now occupy.

There seems to be a matching of these movements by the numbers of sea-floor-dwelling creatures whose shells and skeletons have been fossilised. In the histories of nine of the ten-or-so different phyla of invertebrates the numbers of different kinds of creature present increased when the continents were separate and diminished as the continents came together.

As Valentine and Moores put it, 'the regulation of species diversity on continental shelves has been accomplished through changes in environmental stability' together with nutrient salts supply and changes in continental configurations. Bigger continents, fewer marine species.

Another factor which must have greatly affected the evolution of life has been the seas' advance onto and withdrawal from the continents. These transgressions and regressions have been very widespread, important and long continued movements, as we have seen. Their durations and times have seemed to lack explanation. But this, too, might be found in the theory of plate tectonics.

The force needed to fragment a continent and spread the pieces is provided by the upwelling of material in the mantle. Where it is not overlain by the continents, it produced the oceanic ridge system. Now the ridges really are huge prominences rising above the general level of the ocean floor, and are as large or larger than the biggest continental mountain ranges. They must contain a lot of rock and they must occupy an appreciable part of the ocean basins. Rough calculations have been made to show that the volume of the ridges and volcanic piles on the ocean floor today is about 2.5×10^8 cubic kilometres, 2,500 million cubic kilometres. If the ridges were somehow to be removed, sea level would go down by 650 metres all over the world. Remove the ridges or the ocean trenches and the level of the ocean is at once affected.

We can imagine that as continents are displaced by the movement of plates they may cause the gradual, if not the complete, destruction of the oceanic trenches where the edge of a plate is conducted back into the mantle. The volume of the ocean basin must be reduced, even if only a little.

Meanwhile, as the continents move apart the new space between them may be occupied by a new oceanic ridge. The water displaced by the advancing edges of the continents cannot entirely creep in around the trailing edge as the ridge occupies much of the new space. The total volume of the ocean basin is reduced, but the water has to go somewhere. It spills over the edges of the basins onto the margins of the continents. A transgression is incurred, providing many new habitats for life in the expanding sea.

Let us take the opposite case now. When continents are being joined together new subduction zones where plate re-enters mantle have to develop. It seems, also, that there is evidence to suggest that if ocean spreading stops, the ridges actually subside. In each case we have an increase in the volume of the ocean basins and the seas withdraw from the continents in an epicontinental regression.

To put all this in a nutshell we can equate transgressions with continental splitting and regressions with continental assembly. From what was said a page or two above, it can be supposed that transgressions would tend to moderate the seasonal climates, while regressions, tending to enlarge the continental areas, may have the opposite effect. So yet another influence upon the evolution of marine organisms is thus attributable to the behaviour of the earth's crust. Not only climate but the actual spread of the waters over those important shelves at the edge of the continents has been controlled by the convection below. No doubt if the oceans had never spread and retreated as they have, the record of life would have been the poorer, the evolution of life less eventful.

Afterthought

Some quite remarkable changes have been wrought upon the face of the earth during the many eons of its existence. The strange procession of oceanic crust from ridge to trench, and the vagrant continents upon the crustal plates are possibly unique within the solar system. So far we know of no other planet that behaves in this way, no other planet to our knowledge has similar surface features and possibly a convecting interior. The surfaces of planets such as Venus and Jupiter remain hidden beneath opaque atmospheres. Mercury has only indistinctly observable surface features, but Mars has plateaux and mountains 15 kilometres high. Once again Mars and Earth are found to have something in common. We do not know what kind of a 'geological' cycle may operate on Mars nor what processes produced such great relief. Mars has only a very small iron core and much of its bulk may be composed of material not unlike earth's mesosphere. Some forms of internal convection might be, or have been, possible. If ever man reaches Mars he will probably look around as soon as he can for signs of crustal movement like that on earth. Perhaps in the small bulk of Mars the heat of radioactivity has already declined to a point where convecting can no longer take place. But in the absence of Martian oceans, the relics of 'plate' generation and movement, if they exist, will make exciting exploration.

CHAPTER X

Palaeozoic Events

THE last few chapters of earth history are much better known than all the earlier ones. Covering a mere 600 million years or so and dotted liberally with fossils, they are occupied by fast-moving geological and biological events. Continents were carried to and fro, mountain chains rose and declined, and not surprisingly, the climates changed locally, if not universally. Different kinds of life appeared, evolved and became extinct. The pageant of living things has been punctuated by lulls or intervals when species dwindled and disappeared in large numbers. The rocks and the fossils together give a more readily understood record of geologically recent events than of the distant past, but even the Palaeozoic era is gloriously illuminated compared with earlier times.

Into this chapter we have to cram accounts of a great many events. What is here is of course only a small part of what actually took place. After all we are dealing with the events of the Palaeozoic era, that is some 375 million years ago ending about 225 million years ago. Although the Mesozoic was only half as long, and the Cainozoic a mere 70 million years, the events happening then have blurred or obscured the testimony that Palaeozoic events left behind. In many respects the Palaeozoic world was an odd, strange place, but in it were established patterns that have continued to the present, and the origins of much that has happened since then may lie largely within the Palaeozoic.

When geologists first drew up a table of geological time as represented by rock formations, they were naturally influenced by the conspicuous changes that occur in the assemblages of fossils present from one layer to the next. In many places in Europe they found that major differences occurred above and below an unconformity. Some of the biggest such changes, in Europe at least, so impressed the pioneers of our science that, as we have seen, they used them to mark off large chunks of geological time without much idea of how much time there would be in terms of years. It was obvious that in the

Palaeozoic era, most of the animals and plants belonged to kinds quite strange to us and most of them are now extinct. The great thicknesses of rock through which they are distributed seemed to indicate a great length of time. Mesozoic fossils more closely resemble living forms and the Mesozoic rocks are not so thick as the Palaeozoic. Hence the Mesozoic era was thought to be shorter than the Palaeozoic and its inhabitants had evolved to (or been created in) a condition nearer that of today's. The Cainozoic was thought to be even shorter in duration with plants and animals rapidly approaching the present biological array. Dating based on radio active minerals has produced dates that would startle some of the pioneers but it has confirmed that the Palaeozoic era was very much the largest of the three.

The geologist in the field cannot do radiometric dating on the spot but he can collect fossils. If he were to hammer his way up a succession of rocks that included Precambrian and Cambrian he could recognise the latter by the presence of indisputable fossils, which most commonly are the segmented little creatures known as trilobites. Many of the earliest Cambrian rocks are coarse, sandy beds (fossils unfortunately are not very well preserved in them) resting upon deformed older rocks below. Later Cambrian and other Palaeozoic rocks appear to be spread over wider areas than these first deposits. As the era drew on, more and more of the Precambrian shields were flooded by shallow seas. Several times during the Palaeozoic era the process was temporarily reversed, but only during the Permian period did a world-wide withdrawal of the seas come about. It's a strange, but as we know, not inexplicable, event in earth history and it makes a crisis in the history of life too. We shall return to it later.

A Continental Paul Jones

The most convenient division of the Palaeozoic era has always been into an early part comprising the Cambrian, Ordovician and Silurian periods and a later part made up of the Devonian, Carboniferous (Mississippian and Pennsylvanian in North America), and Permian periods. When these divisions were first proposed it was not thought that they might fall into spans of time characterised by the particular movements of continents or the development of geosynclines. In fact continental drift was then unheard of. As it happens, during the early Palaeozoic era land masses may have disengaged themselves from a continental conglomeration, a super-continent, which has been called Pangaea 1, only to return to each other's company at the end of the early Palaeozoic. In the late Palaeozoic era other errant segments of Pangaea 1 were to return to the fold (producing Pangaea 2).

First let us take a look at how things stood between the oceans and the continents at the very beginning of the Palaeozoic. Early Palaeozoic time, from early Cambrian to late Silurian, covers about 275 million years and as the earth sailed through it she acquired a few scars that are still very much in evidence. To begin with one should emphasise that our view that the cratons or shields were probably closely assembled if not actually in contact as a single cluster is based on flimsy evidence. Palaeo-magnetism yields some suggestions as to where the poles were relative to the continents, and if we believe the magnetic poles always to have been near the rotational poles we can postulate the sites of the ancient continents and the kind of drift that may have occurred.

From this and other evidence it seems Pangaea 1 lay predominantly in the southern hemisphere with the Gondwana continents grouped tightly together and twisted round through about 180 degrees and 'North America' and 'Europe' strung out a little to the east of South America and west of Angara (north-east Asia). The lines along which the continents were ultimately to part may have begun to appear at this early date, with faults giving rise to scarps and valleys between the shields. Much of the terrain was of crystalline rock overlain here and there by thick spreads of Proterozoic sediments, including the recent glacial and fluvial deposits. Large areas of the old shields may have been reduced to lowlands by long ages of erosion. No trace of land plants was to be seen. The ground was barren, much of it baked by the suns of summer and frozen in winter.

As earth passed into the Cambrian period, it may have been a strangely quiet planet, compared with its previous violently volcanic and orogenic past. The ice age just endured had left its marks upon Africa, Australia, Greenland, North America and Europe. The land was perhaps still, untroubled by movements for a while, except in a region near the equator of those times where Australia lay. No doubt within the continental cluster there were signs of the disruption, fuss and fireworks to come. At the time, however, the most significant changes to the land seemed to involve the beginning of a major phase when the seas spread over more and more of the continents.

What we shall never know is how matters were within the ocean basins. The Palaeozoic ocean floors have long ago gone down the sinks, replaced by material added in Mesozoic and Cainozoic times. Nevertheless it is quite possible that there were growing ocean ridges, growing in fact so vigorously that the capacity of the ocean basin was being progressively diminished. The water spilled over the edges on to the continents.

Before the Cambrian period was ended Pangaea 1 began to come apart at the seams, perhaps where the ocean ridges began to extend landwards and lever apart the continental slabs. By Ordovician time large fragments were drifting away from the rest. In North America, Europe and on the margins of Asia violent volcanic activity broke out and earth movements grew in size and frequency along the eastern margin of North America and the facing coasts of Europe. Climatic changes occurred too, though these were probably slow and only of local significance.

As the continent-bearing plates of the earth's crust began to move, seas rose and fell, so that by the end of the early Palaeozoic new mountains were rising from the old geosynclinal areas of ocean floor between Europe and North America. Across wide areas of the Precambrian basement, now barely awash in tropical seas, the activities of marine life left great sheets of limestone.

But we anticipate events. Before the end of the Cambrian period the Canadian, European and Angara Shields were tending to float into relative isolation. The seas opening between them became wider; in Ordovician times the seas became truly oceanic, with new basalt floors. As the northern blocks drew apart from the southern continental group oceanic ridges rose between them and the sea level rose too. The great marine transgressions of the Cambrian resulted in sand and silt from the continental surface being deposited as great sheets all around the margins and even some of the heartlands of the continents. These seas must have offered plenty of colonisable environments for primitive marine life. Plants, the marine algae, would have settled on the bottom and animal life would move in with them.

Towards the end of Cambrian times there were shallow seas throughout much of the equatorial region. Lime-secreting organisms, plant and animal, flourished and great volumes of carbonate mud accumulated over much of the previously sandy sea floor.

As time wore on into the Ordovician period the drift of the northern continents allowed the development of new features on the ocean floors between them. Once the width of these northern oceans had become wider than the mid-oceanic rises and ridges there was more space available for the water. In response the strandline retreated to lower levels round the continents. The regressions did not last long. The early Palaeozoic northern oceans probably never opened very far and the movement of the plates must have been somewhat irregular. Now perhaps they slid in one direction, now in another, depending upon the relative movements of the plates within which they were set.

From palaeo-magnetic evidence and fossils the inference is drawn that the Ordovician equator ran across North America from California to northern Greenland, across the British Isles and over central Europe. Far to the east Angaraland also lay across the equator. The continents were probably as a whole moving in rather an odd way across the face of the earth. Antarctica was moving in a curve south-eastwards from the equator while Africa and South America were also swinging clockwise with them to bring their present north coasts somewhere around the 30 degrees south latitude.

During the Ordovician period there was fierce volcanic activity in several parts of the world. As we have already noted, along the eastern margin of the Canadian Shield and the western margin of the Baltic Shield, for example, piles of basic and intermediate lavas and volcanic ashes were thrown up. There were local volcanic centres here and there not far from the shoreline. Many were explosive sites, coughing out new material in great quantity for millions of years. Some such centres were very large indeed. During quiet times life moved into the warm waters around the volcanic edges of the land but mass exterminations occurred from time to time when life was poisoned by the volcanic gases or smothered by falls of ash.

By this time, it seems, the opposing edges of the Canadian and European continental plates had become destructive. Deeper and trenches were developing parallel to the shores and the volcanic activity must have been increased, as it had been perhaps initiated, by the heat and energy generated along the margin where oceanic plate edged under each continent.

Undoubtedly the most closely studied of all early Palaeozoic geosynclines is this region between the Canadian and Baltic Shields. It stretched from the northern margins of these basement blocks through what is now western Scandinavia, east Greenland, and Britain into the Appalachians and on into south-eastern U.S.A. Known as the Caledonian geosyncline in Europe and the Appalachian in North America, it became the site of some of the classic studies in geology and has revealed most of the sedimentary, volcanic and other characteristics of geosynclines. Its gradual closure during the time between mid-Ordovician and Devonian brought about a series of orogenies that had drastic effects in a region stretching from what is now Spitsbergen in the north as far south as New York. Its climax in Siluro-Devonian times is known as the Caledonian orogeny. At this time the western and central parts of Laurasia were brought together into a clinch that lasted until late in the Jurassic period when the Atlantic rift began.

On the far side of the North American plate another destructive margin seems to have run from California to Alaska with a branch continuing eastwards to northern Greenland. Here it most probably linked with a great mobile belt that stretched from the Russian Arctic south along the site of the present Ural Mountains almost as far as the Caspian Sea and eastwards to central Mongolia. This of course was in part the site of the ocean branch separating the Baltic and Angara Shields. Although the Caledonian orogeny sutured together the eastern Canadian–Greenland and British–Scandinavian areas, the Ural sea remained open. On the ocean-side of the western Canadian (destructive) plate margin an open ocean prevailed.

So much for the unruly north. In the southern half of the Palaeozoic super-continent a rather different train of events had been taking place. There were several regions, notably in Australia, that had received thick Eocambrian sediments. Elsewhere there were complex belts that had been deformed just before Cambrian times. In their present scattered positions, the southern continents show little wrapping of Palaeozoic or later orogenic belts around Precambrian shields as in the northern hemisphere. For hundreds of kilometres their coastlines are made of Precambrian rocks. Behind the coastlines the interiors of the continents reveal for the most part late Palaeozoic and early Mesozoic rocks that are essentially continental, non-marine in origin. They provide us with an important clue as to what was happening on land, what plants and creatures lived there, during nearly 200 million years. An especially interesting view of this kind of situation is provided by the Karroo system with its coals and reptilian fossils. Our chapter on continental drift rather stressed the similarities in the geological records and histories of all the southern continents, and the unity of development that continued into Cretaceous time. Moreover, these continents seem to have had little close connection with their northern counterparts early in the Palaeozoic era. An ocean freeway which was the forerunner of Tethys kept them separated in the east and the forerunner of the Caribbean in the west throughout much of these times.

Not very much is known in detail about the disposition of lands and seas over the present southern continental blocks during the early

Palaeozoic era. Long exposure to erosion since then may have removed some of the evidence, more may be buried under later strata. However, we begin our survey of Palaeozoic events in the south by recalling that the Adelaide Series of Southern Australia is a group of sandy rocks laid down over a long period of pre-Palaeozoic and Cambrian time. Here was a gently but persistently subsiding basin, fed by sediments washed from the low ground not far away. It existed

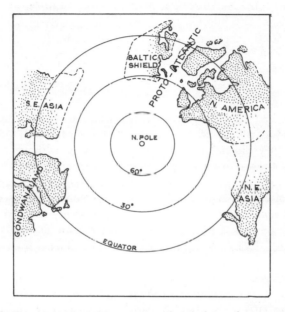

Figure 10-1   A north polar ocean existed throughout much of the early Palaeozoic era. Here is a reconstruction of late Ordovician world geography by H. B. Whittington and C. P. Hughes which incorporates palaeomagnetic and palaeontological evidence for the positions of the continents.

for 100 million years or so, i.e. until the late Cambrian when there was a full-scale orogeny. Nothing so rigorous was in progress anywhere else in the world. No late Cambrian mountain-building episodes are known elsewhere. The rest of Australia, like the rest of the southern hemisphere, has little record of Cambrian geography.

Ordovician rocks are known in few areas but there is the remarkable evidence of an Ordovician glacial period in the Sahara area.

Probably this part of Africa was then sufficiently far from the equatorial zone, fed by snow and high enough to preserve ice for several million years. It is not easy to understand how or why such a glaciation arose but the evidence for one is undeniable.

The Silurian period continued the spread of very shallow seas over much of northern and western South America and north-western Africa and parts of Arabia. The sediments deposited in them were muddy and sandy for the most part, not at all like the great carbonates of the North American and Russian cratons.

Ordovician and Silurian rocks with rather poorly preserved fossils occur in South America and in South Africa but they do not tell us much beyond the fact that for brief intervals there were shallow seas in those regions.

Second Round: The North

The later half of the Palaeozoic era seems to have witnessed the reassembly of all the errant continents into Pangaea 2. Like unruly children, they did not all come immediately or without demur, but eventually the group was complete and a super-continent was formed. A single continent of such size must have brought about appreciable changes of climate and among the late Palaeozoic rock associations we might notice that red continental deposits, not to mention desert accumulations such as dune sands and evaporites, are thick and widespread. Every continent has its late Palaeozoic red rock sequence.

The first signs of the approach of these conditions comes in the Devonian. The Caledonian orogeny in the present North Atlantic region had effectively coupled the Canadian and European Shields and in the mid-Devonian earth-movements the process continued along the margin as far south as New York. Because they mostly affected the area inhabited by the Acadians, the early French settlers of maritime Canada, these are called the Acadian earth movements. Much of the new upland country produced by these orogenies was soon reduced by erosion from bleak and barren hills and mountainsides to more rounded terrain. Between the hills local wadis, lakes and alluvial flats soon developed. In some regions there were extensive sheets of water, lakes in which plant life and early fish were abundant. Around the margins of some of these there were wide shorelines and banks of wind-blown sand. Towards the end of the Devonian a green mantle of vegetation had spread over much of the lowland, scanty and thin in some regions, bushy and thick in others.

The land area that arose in the North Atlantic region has been called North Atlantis or the Old Red Sandstone continent. It spanned what is now the North Atlantic but perhaps the lines along which it would break in the Mesozoic were already established. Certainly there were long deep faults in the crust here and great basaltic and andesitic volcanoes, and during the rest of the Palaeozoic era it was progressively eroded away and covered by sediments. Torrential rains, high

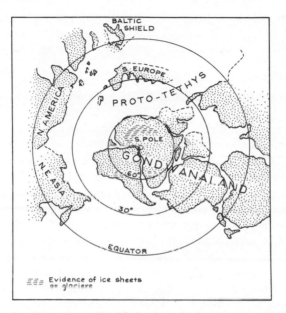

Figure 10–2   The other side of the late Ordovician world from that shown in 10–1 presents us with an antarctic continent. This arrange ment had long been suspected but only recently have glacial deposits been found confirming the cold climate of those days. This reconstruction is also by Whittington and Hughes.

winds and temperatures would have done great damage to the land surface year by year, reducing it eventually to low monotonous relief.

In the other parts of the world the same process was taking place. The Angara Shield suffered from a smothering by debris eroded from itself. It was a world of red and yellow lands in blue and green seas.

Meanwhile there remained open an oceanic route between the Angara Shield and North Atlantis and another still existed separating North Atlantis from Gondwanaland. While the former remained open until well into the Permian, the Variscan geosyncline was to develop from the latter during the Carboniferous period. The inevitable end of the Variscan was an orogeny which affected a region stretching from the southern and eastern U.S.A., where it is known as the Appalachian orogeny, across part of the belt already compressed by the Caledonian upheaval to southern Britain, to southern Europe and eastwards to the Altai Mountains in Russia.

The shallower Variscan seas were perhaps rather like those of the Caribbean today, warm, richly populated by corals, shellfish and other creatures. Here and there were ridges and island arcs with volcanoes. In deeper regions, stretching from the eastern part of Canada across southernmost Britain and southern Europe, thick masses of sediment accumulated further from shore and in situations which suggest the destructive edge of a crustal plate.

The situation changed throughout mid and late Carboniferous times as a series of earth-ripples or orogenic spasms ran through this belt between the northern continents and the southern. New upland ridges were formed forcing up the warm moist air masses to cool and drench the land in rain. In the hot, wet climate their rocks were rapidly weathered and eroded. Over the ground there was now a green coverlet of vegetation, thick in the wetter regions, thinner in the drier. From the new highlands streams and rivers swept down great quantities of sand, gravel and mud. In the lower areas and on the coastline this debris came to rest, locally forming great deltas, coastal plains and swamps. These areas and the large inland lakes that had spread over many parts of what is now western Siberia, Europe and eastern and central North America, were to become the sites of the most prodigious forest growths of all time. The great Coal Measure forests were at their best along the steamy coastal regions and they persisted despite repeated inundations by the sea and burial under floods of sand and mud from the hinterlands. Various modern swamps such as those of Florida and Louisiana are thought to be the modern equivalents of the late Palaeozoic coal forest swamps.

There were forests in many parts of the world during the Carboniferous period. In North America they covered about 260,000 square kilometres of the mid-continent region; in Europe perhaps 100,000 square kilometres. There is a distinct difference between the assemblages of plants in North America, Europe, North Africa and Asia and those in the southern continents, Gondwanaland. While the European floras were tropical, the *Glossopteris* 'fern' flora named after a prominent and typical fossil plant was adapted to temperate conditions, and flourished across enormous areas of the lowlands.

Then at the end of Carboniferous time and early in the Permian period there came the final great episodes of Palaeozoic earth movements. They were to be felt in the Appalachian area stretching into southern and central Europe, and also in the extreme north of Canada extending into the Urals and all round the Angara craton.

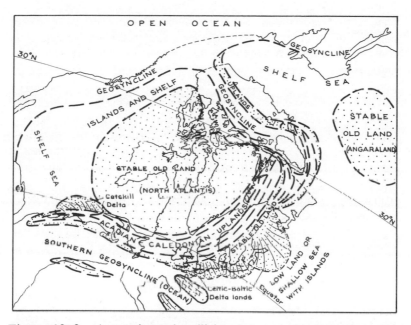

Figure 10-3 A continental collision between the Canadian and Northern European Shields brought about the Caledonian upheaval and the Acadian–Caledonian uplands, the area known sometimes as the Old Red Sandstone Continent or North Atlantis.

The continental fragments were gathering for the great Pangaea 1 reunion. Laurasia was completed as the last real geosynclinal areas were squeezed out of existence. The Variscan and Ural belts were crushed, great alterations and deformation occurred in the central and deeper lying parts of them. Large plutons were intruded into them from below, and volcanoes broke out at the surface. There were similar events in West Africa.

The Appalachian orogeny also involved much crushing, folding and thrusting of the rocks in eastern North America. It seems to have been largely concentrated into the Permian period in North America. The fierce volcanic activity widespread in Europe was not extended into the west.

The result of these upheavals was that essentially all of Europe and North America became land. In central Europe and parts of Russia, in the high Arctic areas of Canada and Siberia and parts of the southern U.S.A. there were limited shallow, very salty seas. These were short-lived affairs, brief dousings of parts of an arid hot continental region. Coral and algal reefs and shell banks sprang up in some parts of the seas, notably in Texas and New Mexico, and in the lagoons to the landward deposits of gypsum and salt were precipitated.

Under such dry conditions the coal forests dwindled and died. The lands were now drier, dustier places with sandy deserts occupying thousands of square kilometres of North America, Europe and central Asia. Sand dunes covered wide areas, wadis were cut in the newly uplifted Carboniferous bedrocks by periodic flash floods and rainstorms. Where gulfs and lagoons of the sea had been cut off from open water they soon dried up. From much of its previous domain the sea withdrew: the lands coalesced to form the great Laurasian continent.

With these events the Palaeozoic era drew to an end, but there was really no lull in the activity of the crust and mantle. In a very short time the plates of the crust were moving in new directions, as we shall see.

Second Round: The South

While the continent-bearing crustal plates of the north were shunting to and fro, the crust to the south seems to have been much less active. In eastern Australia, it is true, a large mobile belt lasted until the Permian period. This, the Tasman geosyncline, experienced many disturbances and volcanic episodes alternating with quiet periods. In South Africa too there was a small mobile belt near the Cape of Good Hope.

For the most part, however, the Devonian period was one in which shallow bodies of water, salty or fresh, covered wide areas of the southern continental shields. Many of them were coral seas, alive with all manner of marine creatures. Elsewhere, as in Australia, there were lakes and streams in which fish and other creatures were plentiful. The land round about supported a simple kind of vegetation. There is little sign of any climatic zonation and fossils of these forms of life have been recovered from several parts of Antarctica. Only in central and eastern South America are there signs that the Devonian had its 'cold spot'. Coarse bouldery rocks there have been interpreted as glacial deposits and the implication is that this area was either mountainous or polar. The south pole may indeed have been situated near the present equator between Africa and South America.

Towards the end of the Devonian period the seas drew back from the Gondwana super-continent and the next chapter in the history of these southern latitudes is a spectacular and remarkable one that has been mentioned before on these pages. It concerns the great Gondwana glaciations and the *Glossopteris* forests. The sediments of the Gondwana ice ages and the coals of the intervening warmer periods are a conspicuous part of the geology of each of the southern continents. In Australia, for example, the sea covered most of the northern part of the continent but from time to time the ice sheets reached out to it from the south. How far the ice travelled is not certain. Along the coastal margins the cool *Glossopteris* swamps and forests stretched for great distances, swept by rainstorms and shrouded in fog, perhaps like the western rainforests of the continents today.

The Gondwana glaciations and the *Glossopteris* forests stretched into what is now eastern India where, again, the ice was moving northwards. In South and East Africa the ice spread northwards as far as Lake Victoria on the present equator. In south-west Africa it was moving westwards. The ice sheets must have been of enormous size, intermittently flowing out in various directions from highlands in south central Africa. Their moraines filled valleys 1,000 metres deep and the ice seems to have reached an arm of the sea in the south, in Madagascar and in Tanzania. There may have been as many as five major glacial ages with warmer spells between.

A similar scene can be painted from South America where widespread moraine was pushed across the old shield surface. Just as in south-west Africa, the moraines choked deep valleys and the ice flowed westward. The first of the late Palaeozoic glacial deposits were found in South America where, as it happens, they are earliest Carboniferous; the African ice appeared soon afterwards. Between the long cold periods, *Glossopteris* forest occupied the well-watered lower regions here as it did on the eastern side of Gondwanaland. So it was also in Antarctica where *Glossopteris* fossils, coal, and glacial deposits have been found in many areas.

The courses of the Gondwana glaciations are not fully known, but the ground from which the ice spread must have been high. There would have been tremendous variations in the amount of ice present from glacial to interglacial phase. When much of the Gondwana ice melted a world-wide rise in sea level would occur, and when the glaciers grew again so much water became locked up on land, and sea level fell. It is tempting to think that this rise and fall could have been responsible for the many world-wide oscillations of sea level in the Carboniferous and Permian periods. These oscillations cut short the lives of the coal swamps around the coasts, the rise of the salt water killed off the trees and plants in the swamps. Unfortunately no world-wide correlation of changing sea levels, coal forests and glaciations in the Permo-Carboniferous period has yet been established. Local earth movements undoubtedly were a contributing factor to the flooding or draining of the swamps.

It all came to an end in Permian times with a progressive drying up of the whole continental area. Ice melted away and streams disappeared. Wide areas of the old shields in Australia and South America were flooded by the shallowest of seas, and when from time to time these were cut off and desiccated, deposits of dolomite, anhydrite and salt were left behind. The ice persisted later in Australia where it stayed till late Permian time.

Pangaea 2

For times older than the Permian the geologist has only a hazy view of the positions of the continents but there is enough evidence, as we have seen, to allow him to construct Pangaea (or Pangaea 2, as we can now call it) on his globe with a fair display of confidence. It is a rather arc-like conglomeration with perhaps a small bite out of the north-western margin and a very much larger one on the eastern side to give the Tethys Ocean. Antarctica has taken up its station in the south polar region. The equator runs from Lower California to New England, North Africa and Spain and out into Tethys. The extreme northern tip of Laurasia lies north of 80 degrees.

Palaeozoic Climate

Looking for evidence of what Palaeozoic climate was like involves several fields of search. There is the obvious area of palaeontology wherein we suggest that fossils resembling tropical organisms today denote tropical organisms in the past. We can use corals, molluscs, vertebrates, plants and a few other fossils in this way. Similarly, the

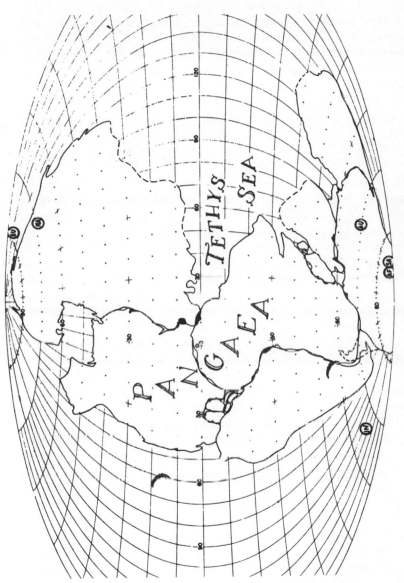

Figure 10–4   A reconstruction of the super-continent Pangaea at the end of Permian time, 225 million years ago.

(From Dietz and Holden by courtesy of R. S. Dietz.)

presence of salt or gypsum deposits in the rocks suggests arid hot conditions like those in which salt pans and salt crust form today. Ancient moraines indicate glacial ice, which in turn demands a cold climate or a mountain terrain. Mountain terrains with glacial deposits do not commonly become part of the geological record. Deserts, too, have distinctive deposits such as dune sands and mineral-cemented soils.

From the Palaeozoic rocks we can draw upon all kinds of evidence such as this, and from palaeomagnetic studies we can determine a region wherein the magnetic poles might have been and what the latitudes of the continents were. Even so, the evidence is not very comprehensive for any one part of the Palaeozoic era. The Carboniferous period has given more recognisable clues than others in the era and the extent of some of its forests, deserts, and glacial regions, as well as the deeps and the shallow seas where reef corals flourished, is drawn with more confidence than for other times.

Beds of rock salt were formed in lagoons and salt lakes during hot, dry spells in the Cambrian. They are also known in the Ordovician strata. Silurian salt deposits on a much larger scale occur in eastern North America, as do very extensive salt beds in the Devonian. The giant Devonian salt deposit that lies beneath the prairies of western Canada and the U.S.A. was described in Chapter VII.

Late Palaeozoic salts have been found in Western Europe, including Britain, in the Moscow basin and the Ural Mountains, the European and Canadian Arctic and other areas. They became so widespread in the rocks of the last few million years of the Palaeozoic that there seems to have been a coincidence of hot climate and a withdrawal of the seas from the continental edges.

The Gondwanaland glacial rocks, the Dwyka and its related formations, are Carboniferous to Permian in age and found on all the southern continents. They have always excited comment, and to account for this ice age there must have been enormous snowfalls, moisture swept up perhaps from the ocean to the south and also from the north and cooled in the high plateau lands in the continental interior. The total area actually glaciated must have been comparable in size to that of the northern hemisphere continents covered by the Pleistocene ice sheets. As many as eleven successive old moraine deposits, one upon another, are known in Australia. In many areas the glaciers reached the sea. Although the Pleistocene ice age lasted 2 million years or a little more, the late Palaeozoic glacial chill may have lasted ten times as long.

Despite the apparent confidence with which geologists draw up maps of the world as it might have been a million years ago, they are mostly aware that they may be hopelessly in error or unaware of vital information in one respect or another. A few years ago no one would have thought that a major glacial event in the early Palaeozoic could have escaped us. But there is always something new out of Africa, and in the course of exploring the central Sahara recently, French geologists came upon Ordovician rocks with signs of glacial origins. Their discoveries were so intriguing that a special expedition was sent soon afterwards to find out more. What they found confirmed the initial report. There are fossil boulder clays or tillites on glacially eroded rock platforms with surfaces striated and polished by the passage of the ice. There are the thinly layered deposits of glacial lakes and the outwash deposits of run-off meltwater streams and rivers, covering huge areas. Palaeomagnetic data give indications that this part of Africa and the northern end of South America were indeed in far southerly latitudes. Cold polar winds would have swept moisture up from the southern ocean into the continental interior.

From other palaeomagnetic sources the evidence points to an Ordovician equator running from California across Newfoundland and western Europe. From tropics to pole then there must have been climatic zones, but how wide tropical, temperate and frigid zones were is another problem. Nor do we know exactly when this arrangement began or ended. We have merely a glimpse of a situation which may have been unique to that part of the early Palaeozoic but which may be repeated elsewhere. The present arrangement of climatic zones of the earth controls the distribution of most living things today. It would be reasonable to think that a similar effect was maintained in previous glaciations. Proving it is not so easy, and we cannot yet see it in the Ordovician.

Where fossil coral reefs occur we can reasonably infer that the water and climate under which they lived were much the same as the tropical climate and clear waters that we know in our own equatorial seas. Fossil coral reefs of a kind first appear in the Ordovician rocks. They are present, too, in the Silurian and Devonian, and Carboniferous rocks of most continents.

Permian reef formations are less widely scattered. Palaeozoic corals together with many of their fellows on the sea floor disappeared from many regions towards the end of the Permian period. By the close of the period virtually all were extinct. The Mesozoic reef builders did not appear until as late as the Jurassic in most parts of the world.

The coal forests that grew between the late Devonian and early Permian periods reveal plants that needed abundant moisture and plenty of sunshine, not to say a tropical climate. Their activity depleted the atmosphere of a great quantity of carbon dioxide and replaced it with free oxygen. The new, more oxygen-rich atmosphere may have had geological significance, speeding the weathering of rocks as well as offering an advantage to the higher animals, the vertebrates.

Life in the Palaeozoic

The appearance of the continents today owes much to the coating of living things scattered on their surfaces. Forests, grasslands and even the Arctic tundra are complex communities of living plants and are the homes and support of many of our more familiar animals. When fossils were first entrapped in the rocks there were no forests, no grasslands and no tundra. Life was for the most part confined to the sea, only a few algae and other very simple plants may have sought an existence beyond the lapping of the waves. Some undoubtedly invaded freshwaters, moving into rivers and lakes, but it was not until half-way through the Palaeozoic era that plants began to colonise the land. Green, buff and russet-coloured patches of vegetation may in late Silurian days have begun to spring up beside the rivers, lakes and other waterways.

They were simple growths, moss-like or slender, reed-like and leafless in appearance, a few inches high and bearing small spore-generating capsules rather than seeds or flowers. Several different kinds of these stood in the shallow waters, bending and swaying like sedge in the wind, and in Devonian time they were joined by the first fern-like plants. A mere 40 million years later the scene had become much more varied with larger lush plants with well-developed leaves and fronds. Real forests of them had taken root by the Middle Devonian and at the end of the period they were reaching 7 metres or more in height, towering over a thick underbrush of ferns, mosses, liverworts and other smaller plants.

Splendid as the late Devonian forests were, they too pale into insignificance beside their Carboniferous descendants. The great coal-forming period of the earth's history originated with the rise of the immense forests of moisture-loving swamp plants towards the end of the Palaeozoic era. Dominating the forests were the tall reed-like forms much like our own horsetail ferns and scouring rushes. Some would be as much as 40 feet high, slender, unbranching, with a thick core of pith and rings of leaves at each node or joint. At the top a cone bore spores to be scattered in the wind. With these plants the scale trees flourished, as much as 40 metres tall and a metre or so around the trunk. Near the top branches sprang out and were covered thickly with needle or splinter-like leaves and tipped with large spore-cones. Other trees had robust trunks and crowns of sword-shaped leaves or many branches with fronds like ferns. The tree ferns grew more than 15 metres high, and amongst the many different kinds of plants found with them are true seed-bearing kinds, more advanced in structure than the ferns.

Such prolific plant growth with its attendant hosts of parasites, fungi and minor plants seems to have required a humid, tropical or temperate climate. From an atmosphere rich in carbon dioxide the growth of the Carboniferous forests may have removed much of it in exchange for oxygen. The carbon dioxide provided carbon for the plants' tissues and in their turn these gave the peat from which coal is ultimately formed. As it rotted, became buried by mud and sand and slowly compressed, the debris of the swamp forests, the peat and decaying vegetation, was turned into the black combustible rocks we know as coal. And the effects of exchanging oxygen for carbon dioxide in the atmosphere? Some geologists have seen the rise of the reptiles as a response to the availability of oxygen and the reduction of carbon dioxide as a reduction of the 'greenhouse effect', as a result of which the climate grew appreciably colder.

If the forests were so big and flourishing, one might ask, what caused them to vanish? Probably it was due to a combination of events. Earth movements may have so altered the configuration of the land that the low-lying swampy areas were drained, or flooded permanently. There were certainly widespread changes of sea level that killed off vast acreages of vegetation. Coupled with these events, the climate may have become more seasonal, locally arid or even desert and in Gondwanaland, as we know and have just mentioned, it took a decidedly polar turn in the Carboniferous. Whatever the causes, the coal forests were drastically reduced in size as the end of the era drew near. Many of their plant species died out.

Forests offer shelter to all kinds of animal life; the Carboniferous forests seem to have been the home of some of the most remarkable fossils associated with the coal plants, including large insects, spiders and scorpions, dragonflies of huge size, and salamanders and other amphibians and fish. Of these vertebrates we shall have more to say anon; for the moment we must return to the sea where the most prolific record of Palaeozoic life was being incorporated in the sea floor sediments.

Life in the Sea

If the beginning of the Cambrian period is thought of as the raising of a theatre curtain, it reveals an already crowded stage. The actors are numerous and diverse; over 900 species of them are known from the Lower Cambrian rocks. They are all marine creatures but they have different roles to play in the Cambrian scene. Even in the earliest of these there are many different kinds of complex, extinct and distant relatives of the modern crustacea such as crabs and the insects, the trilobites. Their ancestry must be a long one, going back into the Precambrian. With them are comparatively few kinds of other fossils and most of them are of animals that lived on the sea floor. We can only speculate about all the kinds of creatures there may have been swimming or floating about at or near the surface. In the sea today our zoologists distinguish those animals that live on the bottom, the *benthos*, from those that inhabit the waters above as *pelagic* creatures, the *plankton* that floats and the *nekton* that swims actively about.

During the Palaeozoic era plankton, nekton and benthos all evolved into many different forms, most of which are now extinct. Planktonic and nektonic animals do not in general have many hard parts that are easily fossilised. Many are extremely small, delicate or fragile. On the other hand, the benthonic animals have invested in shells, carapaces, and other kinds of resistant structures as protection from predators or a turbulent environment. They are more robust and compact. Their evolution is well documented by countless millions of fossils.

The Cambrian sea floor was peopled with a great variety of trilobites. Nine-tenths of all the Cambrian fossils known are trilobites. They were segmented animals a few centimetres long, with distinct heads, bodies and tails. Many were spiny or covered in tiny wart-like ornamentation. They had paired legs beneath the body and antennae at the head. The mouth was on the underside of the head while a pair of eyes was perched on top. Some of the later trilobites had eyes on stalks, others developed compound lenses like a fly's, yet others seem to have lost eyes altogether. As the trilobite grew it discarded its old skin or case by splitting it at the head and emerging as a larger individual. Trilobites may well have been scavengers and mudgrubbers, some living in shallow waters and some in deep. During this long evolution they produced many different forms; some of them were very bizarre. They had their greatest diversity in the Cambrian, but almost all of the five great divisions or super-families of them alive during this period died out at the end of it. Trilobites steadily declined in numbers from then onwards. Only a few species were alive in the Permian period and none are known from later rocks. The disappearance of the trilobites may have been due to steadily increasing competition for existence from the multiplying other animal groups in the seas. Nevertheless the trilobites were around for 350 million years or more, so it cannot be said that they gave up the struggle easily.

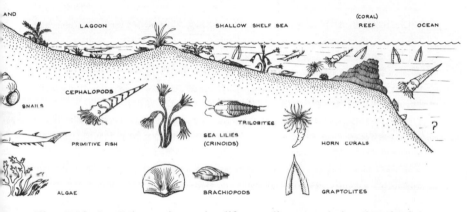

Figure 10-5   Palaeozoic marine life was diverse and abundant in the shallow seas at the edges of the land. Many strange forms were present and this diagram shows a few that may have been present towards the end of the Silurian period. The sketches are not all to the same scale.

In very many areas companions to the trilobites were the small-shelled bivalve creatures known as brachiopods. They are still alive today, though in much depleted numbers compared with the Palaeozoic. There are brachiopods with two small spoon-shaped or circular shells of horny material and brachiopods with two more robust shells of calcium carbonate. Although the brachiopod larvae swim around freely, the adult is firmly anchored in mud or sand or to a hard object on the bottom. Cambrian brachiopods were mostly of the horny kind. Later the lime-shelled forms evolved into a truly remarkable variety of shapes, and some became as big as coconuts. They seem to have preferred shallow seas, warm and free from mud, and like the trilobites they never ventured out of the sea. Their heyday was undoubtedly the Devonian period when they occupied the sea floor in amazing numbers. At the end of the Palaeozoic era the brachiopods, which had been so prolific the world over, shrank to a mere handful of varieties, a few of which still remain. The brachiopods may be said to have had a mid-Palaeozoic success, and like the trilobites, they may have surrendered largely to the competition put up by the new animals, principally the molluscs.

Nearly all the shellfish today are molluscs of one sort or another, descended from simple, perhaps segmented, ancestors living on the sea floor. Snails, clams, cockles, mussels, cuttlefish and even the squid and octopus are members of this group. They are soft-bodied but manage to secrete one or more shells for protection. Molluscs have had a successful history and seem to go from strength to strength. They have swum, crawled or glided into almost every possible environment, and they come first to our attention late in the Cambrian. From these small animals arose in Ordovician time snails, clams and many kinds of squid-like creatures in cone-shaped shells up to a metre or two in length. In the later Palaeozoic periods these all flourished in many parts of the world. The Palaeozoic seas presented few barriers to these hardy and buoyant animals: they have left their shells on every continent. The larger squid-like forms became extinct but their smaller coiled relatives left descendants to enter the Mesozoic era.

Among the stationary animals to colonise the sea floor, the sponges and corals were both much in evidence. There were other animals, known as archaeocyathids, that seem to have been a cross between the two and to have had a cone-shaped housing of calcium carbonate 10–20 centimetres high. These creatures grew together in clusters that were quite like mats or thickets of small corals in California, Newfoundland, Australia and Russia. The sponges never seem to have amounted to much of an important group of marine invertebrates, having advanced but little along the path of evolution since the Cambrian period.

The stone corals, on the other hand, or at least several extinct branches of that group, put in an appearance in mid-Ordovician time and, in one form or another, spread rapidly throughout the shallow Palaeozoic seas to form banks, reefs and fields much as corals do in the warmer shallow waters of the oceans today. Together with lime-secreting algae and, locally, the colonies of calcareous bryozoa or moss-animals, they produced a completely new kind of sea floor. It had nooks and crannies offering protection to other animals. It was well aerated and oxygen was locally at relatively high concentrations because of the photosynthetic action of algae. Oxygen provided a basis for increased metabolic rates in many animals and thus perhaps aided evolutionary processes. There was an abundance of calcium carbonate either in solution or as debris from broken coral skeletons which other creatures could use in their own skeletons or shells. Some of the coral banks or reefs were steep-sided ramparts rising scores of feet above the sea floor, and some of them may have existed for many millions of years before being finally buried by sediment or lifted above sea level.

Just as modern tropical seas may swarm with starfish, sea urchins and sea lilies, so the Palaeozoic seas had their communities of sea lilies and other members of the phylum echinodermata, the 'spiny-skinned'. In the Cambrian seas there were many different classes of echinodermata but by the middle of the era it was the crinoids, the delicate, stalked, flower-like group, that were the most common. Over vast areas of the Carboniferous sea floor they lived by the millions, raising their fragile calyces as much as a metre from the bottom.

One might have seen all these animals in a quick glimpse of the sea floor in, say, Silurian days. But of course there were also many other kinds, creeping, gliding or simply just staying put on the bottom, creatures that have left no hard parts to become fossil. There were protozoans, tiny single cells of protoplasm, by the million. Only when they, too, developed a hard case of calcium carbonate late in the Devonian period did they bequeath something for the fossil record. When this happened the remains of these tiny animals accumulated in countless numbers to give us, as the corals and the crinoids did, thick blankets of sediment on the sea floor. It became in due course the limestone of today.

So much for the sea floor, but the waters above were just as full of life. Floating or swimming creatures, feeding on the planktonic plants, had only light shells or skeletons for the most part, just as their descendants have now. Many of them had no skeleton at all, jelly-fish for example; others, such as graptolites, had a light horny casing to house elongate or radiating colonies of minute individual creatures. There were plenty of these organisms in the waters of the late Cambrian, Ordovician and Silurian seas, but together with other shallow-water planktonic forms of life they became extinct in the

Devonian; for what reason we may never know. Only in the last few decades have these fossils, rarely more than a few inches long, been found preserved in full relief. Most occur as streaks of flattened carbonaceous matter in the fine-grained rocks. The architecture of the graptolites is puzzling but in some details it resembles that of creatures allied to the most primitive vertebrates, which is a strange and unexpected relationship.

So abundant were graptolites on occasion that they have been pictured as forming Sargasso Sea-like areas of crowded floating fronds and colonies. One can imagine them rising and sinking in the water as they attempted to follow the smaller plankton for food and moving with the ocean currents across the face of the earth.

There are many other puzzling and strange fossils known from the pelagic communities of the Palaeozoic rocks. Some are rare, others common, most are small, even microscopic. No doubt many of them formed the lower steps of the ecological pyramid of the day, the successively higher steps of which consisted of animals that fed upon those of the step beneath. At the top would have been the great predators—the swimming molluscs and, of course, the fishes.

Fishes are truly masters of the realms of water, yet we know little about their origins. Fishes are vertebrates, animals of complicated structure with skeletons of bone or cartilage and in the course of perhaps 400 years of evolution they have swum into every kind of aquatic environment. At some early date in the Palaeozoic there must have been somewhere in the sea the ancestors of the vertebrates, but since they had not evolved any hard tissues that could be fossilised we have not identified them for certain. Argument as to which group gave rise to the first vertebrate has always been rather energetic, so too has been the argument about whether or not the sea rather than fresh water was their place of origin. By late Silurian times some of the earliest and most primitive vertebrates were indeed poking about in fresh waters. Fresh water, brackish water, foetid and muddy water and even the cold dark water of the oceanic trenches all have their specialised kinds of fish today. From this versatile and prolific group came the first air-breathing, four-limbed creatures, amphibia. From amphibia the reptiles evolved, and from reptiles were derived the birds and the mammals. The chain of evolution from fish to man from early Palaeozoic to the present has many missing links and its beginning is lost, perhaps for ever. The oldest pieces of bony material occur in middle Ordovician marine rocks in America. Simple jawless marine vertebrates make an appearance in the Silurian scene, but in the

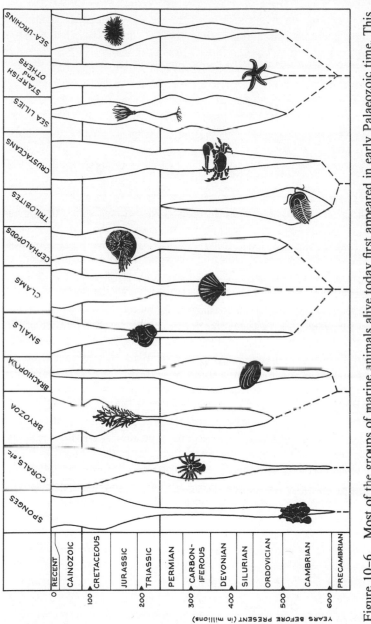

Figure 10–6 Most of the groups of marine animals alive today first appeared in early Palaeozoic time. This figure shows how the variety and numbers of these creatures have changed over the years, the width of each column corresponding to abundance.

Devonian period there was a veritable explosion of the scaled and finny. Perhaps the rivers and lakes of the new Devonian continents became accessible at a time when the fish had reached a point in evolution where they could adapt to non-salty waters. Adapt they did, and soon gave rise to a host of different kinds, some flat, fat and sluggish, others streamlined and active. Many of those we know best were encased in bony plates of armour. Bony fish abounded in lakes and rivers and must have spread widely in the seas. The sharks, primitive predators with cartilaginous skeletons rather than bone, sidled and glided through the Devonian seas. Most spectacular of all fish were the dinichthyids (or 'terrible fish'); 8 metres or so long and with the huge jaws and slicing teeth of predators, they lived in the late Devonian seas covering North America and North Africa.

Those hardy creatures the lung-fish, with their ability to survive periods of drought and to breathe air, made their debut in the Devonian period too. By late Devonian time they were accompanied by other bony fish that not only undoubtedly had lungs but also had stumpy or lobed fins, the antecedents of legs. From water to land, at least for short spells at a time to begin with, these fish, 30 centimetres or so long, had made one of the most significant steps in all of organic evolution. This was no conscious move into the terrestrial way of life, perhaps, so much as a slow, brief plod across the mud flat between one pool that was stagnant or drying up and another that was not, but it began a process that led to a more adventurous existence. To return to the water was always necessary but for some it was eventually only to lay eggs there. There were new realms to be explored above the water-line by the new arrivals and they acquired the anatomy and habits that mark them as amphibians. The newts, salamanders and frogs are all that remain today of a group that became highly successful and varied in the Carboniferous and Permian periods. Coal swamps undoubtedly would have provided favoured habitats for them. Some of the amphibians there grew to 2 to 3 metres in length: some were slender, others rotund, probably moving only sluggishly; all were obviously flesh-eaters with powerful teeth and jaws.

From fish to amphibian was no mean evolutionary achievement, but the late Palaeozoic era saw yet another, equally important. Some of the four-legged kind managed to reproduce without returning to the water. By evolving eggs which allowed the animal to develop through a larval or tadpole stage to a stage where it resembled a fully grown adult before hatching, freedom from the watery environment was gained. This kind of egg, the amniote egg, has been hailed as the most marvellous 'invention' in vertebrate history. Together with this went several changes to the skeleton and no doubt other parts of the body, and the more active and versatile creature that resulted was, in short, a reptile, the highlight of the Palaeozoic animal world. The rise of the insects in the Carboniferous provided a generous basic food supply for the early land dwellers. In what might geologically be called no time at all there were reptiles of many different kinds in several parts of the world. Some were tiny, others the size of cattle. Herbivores and carnivores, they lived in many different kinds of territory from swamp to semi-arid land. A few even took up life again in the water.

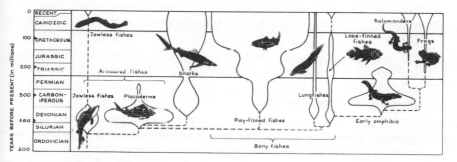

Figure 10–7   The mid-Palaeozoic was a time of great change and expansion for the fishes and other backboned animals. Most of our modern great groups of fish and amphibia have their origins at that time. Only in the late Palaeozoic did the reptiles appear. The width of each column in this diagram represents the variety and abundance of the group named. (After A. L. McAleseter.)

The End of the Palaeozoic

We may imagine that as the world began drastically to change its geography in the Permian period, the reptiles sought to occupy the different habitats that appeared. Geographical changes and the bio-logical consequences of these would affect all classes of life and the Permian period brings us to a time of real crisis for living things at the end of the Palaeozoic era. Extinction overtook many different groups of animals, terrestrial and marine, vertebrate and invertebrate alike. How is it to be explained from what we know of the record?

The Permian period was certainly one in which great changes to the continents were taking place. Pangaea may have been undergoing warping and fracturing. There was violent volcanism in several parts of the globe and the sea seems to have withdrawn from the continental areas, rather as if the latter were being gently raised from below. Climatic changes with a protracted glacial epoch in Gondwanaland and, to the north, conspicuous seasonal changes in more temperate regions were taking place. Few people believe that these events alone could account for the extinction of so many different kinds of corals, brachiopods, echinoderms, arthropods, molluscs, reptiles, bryozoa and foraminifera, to mention only the creatures well-known as fossils.

Cosmic radiation, the explosion of a supernova far out in the universe, and a sudden burst of solar radiation have both been suggested as possible causes. The lethal cosmic rays may have penetrated the atmosphere and had a drastic impact upon living organisms without perhaps having any effect whatever upon climate. The snag here is that the land plants, which would have been the most exposed to such a dose of radiation, seem to have been little affected by any changes while the animal world was being so decimated.

Locally the extinction of one group of animals can cause that of other groups dependant upon it for food. The effects of this could eventually be quite far-reaching and penetrate the inland habitats far from an inshore marine region where a particular key extinction began. The combination of the withdrawal of the sea from the edges of the continents, coupled perhaps with some climatic variation, might have initiated such a wave of misfortunes. As an idea this has much in its favour but trying to prove it actually happened that way is another matter altogether and beyond our means at present.

Climax and Death

So what have geologists been able to conclude about the evolution of the earth and its inhabitants during the Palaeozoic 370 million years or so of its existence? It seems to have been an era of great change and activity that proceeded slowly at first but accelerated towards the end of the era. The shields were flooded by and drained of shallow seas from time to time. Geosynclines with volcanoes and, inevitably, orogenic death-throes, appeared alongside many of the shields. Mountain chains were formed and eroded away. It all has a familiar ring. The geological processes seem to have been in keeping with those observable today and there is no obvious reason why continental drifting as a result of crustal plate evolution should not explain it all. Well, there may be no obvious reason, but the actual collection and organisation of enough information to put it beyond doubt remains to be achieved. How in detail the continents moved together to form Pangaea—assuming that they actually did—is not proved.

The climatic changes of the Palaeozoic may well have been produced by changes in the distribution of land and sea. They may also reflect changes in solar radiation and changes in atmospheric composition. Eocambrian glaciations, an Ordovician ice age and a Permo-Carboniferous glacial period constitute a fairly substantial record of the lows on the Palaeozoic thermometer. It is doubtful if the changes in the fossil record can be related directly to these refrigerations, but they must have had serious effects locally and temporarily.

Similarly, the seeming aridity of much of the continents towards the end of the Palaeozoic must have cast an influence upon the evolution of the terrestrial floras and faunas. It was a state that continued well into the Triassic period in many parts of the world. If it did indeed affect land animals and plants there is no great wonder, but the extinction of so many marine organisms towards the end of the Palaeozoic is something quite different. So far there is no completely satisfactory answer to the question of why it happened.

One attractive idea we have mentioned before is that with the reconstruction of the Pangaean super-continent an extreme condition was produced. The marine continental shelf areas were reduced to a minimum and competition between different groups for the available space and food would soon eliminate many species. Large seasonal changes in climate and in the nutrient salt content of the nearshore seas might have resulted. This could have sounded the knell for many organisms, beginning with plants and working up the ecological pyramid.

The sum of all these changes, wrought upon the world by the end of the Permian period, however, were presumably enough to put 'finis' to the record of many kinds of life. They were also changes that produced new environments for the organisms hardy and adaptive enough to enter them. Those animals and plants that did survive or evolve to meet the new challenges were the founders of the Mesozoic communities that were soon to develop.

CHAPTER XI

Mesozoic Affairs

MESOZOIC rocks cover large areas of the continents today. They lie contorted in the great Cordilleran ranges of the Americas and in the Alpine and Himalayan mountains: they spread across the great plains of the American west and over wide regions of the African veldt and the Russian steppes. New Zealand has a mountain backbone of Mesozoic rocks and there are mountains built of Mesozoic geosynclinal formations in Antarctica. Most of these rocks are sedimentary but there are also great piles of volcanic lavas and ashes in parts of western North America and around the Red Sea, to name but two areas where they occur in the Jurassic. The vast outcrops of Mesozoic flood basalts in South Africa point to volcanic activity on a colossal scale in that region too. On a smaller scale, volcanics in Australia, Antarctica and the North Atlantic burst out at various stages in the Mesozoic era.

Reconstructing the geography of a period in the Mesozoic is rather easier than for one in the Palaeozoic. The rocks on which to base our interpretation are more widespread, less covered by later deposits. We can distinguish easily the land areas, the wide shelf areas flooded by Mesozoic seas from time to time, and the Alpine-Himalayan and Circum-Pacific geosynclines. From the dawn of the Mesozoic era onwards world geography has steadily evolved from a relatively simple affair to a complex arrangement of widely different and well-separated continents. Animal and plant life has responded to this and the floras and faunas of the continents and their coastal waters have become more different from one another as time has gone on.

The starting point for the history of the Mesozoic must obviously be Pangaea 2. Pangaea 2 may not have been a very tidy grouping of the continents but at least it meant that there was fairly easy access between them for many different kinds of animals and plants. During the Mesozoic era this changed very much with the Americas separating from one another and the Gondwana continents moving apart and northwards. The great Tethyan bight was largely to close and within this region new island arcs and uplands would rise. The Atlantic and Indian Oceans would be growing towards their present sizes and shapes. All this was to have a profound effect upon both marine and terrestrial life. Communities on land and in shallow inshore waters would become cut off from their neighbours of long standing. For each new continent new provincial or regional characteristics developed, influenced by the changing conditions and especially by increasing isolation, and perhaps by new patterns of ocean and atmospheric circulation.

When the Mesozoic era began, the earth travelled round the sun at the same rate as always but the length of the day was somewhat less than 24 hours. There were 385 days in the year; when the era drew to a close the number had declined to 371. What difference, if any, this could have made to the rhythm of life in those times is unknown, but it would not have been very significant. On the other hand, the migrations of the continents relative to one another and to the magnetic poles might have had appreciable effects upon migratory or planktonic organisms. The geological reasons for the evolution of some animal migrational behaviour almost certainly lie in the late Mesozoic.

Pangaea Splits
Whatever and whenever the forces were that drew the continental blocks together to form Pangaea 2, they seem to have been reversed from Permian time onwards. In a span of about 45 million years, that is by the end of the Triassic period, rifts had developed to separate Mexico from the northern edge of South America and the east coast of America from north-west Africa. Northwards from New England Laurasia was still intact, though it was undoubtedly creaking a bit. Faults and the spilling of a basaltic lava took place here and there and sundry warpings of the surface occurred from time to time. This may well have been a reaction to the tendency of western Laurasia to swing north-westwards as the North Atlantic rift and the Mexican-Venezuela straits opened. To compensate for at least some of the sea floor spreading some of the local crust had sunken along the centre of Tethys; it had been driven down or pulled into the Tethyan Trench.

South America and Africa-Arabia still stuck together but a long bight developed to separate western Antarctica from the southern tips of South America and Africa up the eastern side of the African–Madagascar–Arabian block towards eastern Tethys. A branch rift ran eastwards from South Africa along the southern (now eastern) side of India towards Australia. The rifts were active volcanic areas from which sea floor spreading took place. The Gondwana continents of Africa, India and South America began twisting slowly in an anticlockwise direction and moving gently northwards. Australia was turning in the opposite direction as was Antarctica. To a greater or lesser extent these patterns of movement have been followed by the southern continents ever since. A sort of slow waltz had begun.

Andante—Triassic

For a view of how things were on and around each of the continents now that Pangaea was disintegrating let us begin in the western parts of what was Laurasia. On the far western margin of North America lay the blue waters of the Pacific Ocean. They flooded slowly eastwards past islands and bars into Alberta and the Rocky Mountain states of the U.S.A. as a long geosynclinal belt subsided from Alaska to Mexico. A shallow arm of this sea swung across the northern Arctic Islands, and northern Greenland and Spitsbergen. True to geosynclinal form, there were volcanoes strung out along parts of the subsiding region of the Cordillera, lavas, breccias, and wind-blown ashes were spread over much of the coastline before the sea crept slowly in.

Between Utah, Arizona and Texas, the land was low and uneven, for much of the time it was desert. Torrents occasionally flowed from the hills to the east, dumping cobbles, boulders, sands and gravels over hundreds of square kilometres. At other times there were shallow lakes or the great rolling dunes of a sandy desert. The dusty winds blew strongly from the west and north-west. All in all it was an arid, hot landscape, although there were scrubby woods or forests here and there with fish, amphibia and reptiles lurking in or near the water.

Eastwards across the continent the land rolled away in endless vistas of dry wasteland as far as Appalachia. This eastern margin of the continent was the scene of crumbling uplands and dribbling volcanoes. The Alleghanian (Appalachian) orogeny had recently completed its compression of eastern North America and the region had now risen well above sea level. Throughout much of the Triassic period and under a baking sun, the ground was raked by erosion and the debris was piled into long narrow rift valleys. There were rivers, ephemeral and braided by islands of gravel but periodically swollen into torrents; swamps and lakes were sometimes invaded by prodigious floods of the turgid muddy water and never far away was the

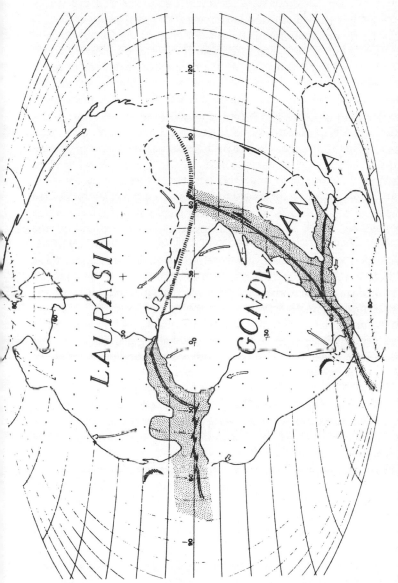

Figure 11–1 The Triassic world of 180 million years ago shows the beginnings of the continental separation that is still going on. The stippled area shows the area of new Triassic ocean floor and the arrows show the directions in which drift was taking place.

(From Dietz and Holden, courtesy of R. S. Dietz.)

possibility of yet more basaltic outpourings from the fault-line fissures. Some of the earliest dinosaurs, including one of the first ever to be scientifically described, strutted and hunted in this part of the world.

Across the developing gulf that was to become the Caribbean and the Atlantic, Europe, too, was largely an arid, hot region. To the south lay Tethys and from Tethys came the rising seas that were to reach across most of central Europe as far north as Denmark during the Triassic and Jurassic periods. The northern Tethyan shore from Spain to the Caspian and east to China lay at about 30 degrees north. The western margin of Europe stood high for the most part, with large regions of Spain, France, the British Isles and Scandinavia providing a rugged bulwark. Wadi-like valleys in the highlands were soon choked with rubble, sand and mud swept into them by the periodic storms, and erosion of the sparsely vegetated countryside proceeded in fierce fits and starts. Erosion and deposition between them smoothed out much of the rough topography by the end of the Triassic period and across the low plains the winds blew clouds of red lateritic dust or swept sand dunes before them. Of vegetation, there was little but scrub and conifers. Between the rounded low hills lay broad shallow lakes, some of which were very salty, bitter to the extent of supporting almost no life at all.

In the mid-Triassic period, however, the sea worked its boundaries across central Europe, to lie like a great shallow puddle among the lands. Warm, clear and shallow, it offered an excellent habitat for shellfish of every description. The German name for the limestones that formed from the shelly sediments is, appropriately, '*Muschel-kalk*'. Within the muschelkalk sea there were many other forms of life. The dolphin-like ichthyosaurs disported themselves in the Triassic seas and along the shores of Tethys there may have been wandering reptiles, alone or in small herds or groups.

Eastern Europe and Asia stretching to north of the Arctic Circle were similarly rather arid landscapes and they were separated from one another by the broad Ural belt where fluvial, lacustrine and shallow marine deposition was going on. At its southern end the low swampy Ural country with perhaps wide and open estuaries met the Tethyan Ocean.

Tethys itself is difficult to reconstruct but its Triassic rocks indicate zones of shallow water, zones of reefs and shell banks, and areas of deeper water. Locally the old shorelines are distinguishable and in the shallow backwaters and lagoons decaying vegetation and driftwood came to rest. There were volcanoes in a few places where we can examine them, and layers of volcanic ash and dust or lava were probably deposited in many other areas.

Tethys swept in a great loop past Burma and Thailand south-east into Indonesia to meet the Pacific geosynclinal belt. On its southern side lay dusty Africa, India and Australia. There were rather vigorous areas of subsidence in the eastern parts of Tethys with many a submarine landslip and mud-slide as sediment settled on slopes not stable enough to bear the load. Thousands of square kilometres of sea floor were covered by sands and silts settling from these disturbances. Volcanic eruptions and earthquake shocks must have helped to set off these cataclysmic submarine events. Having reached the far eastern end of Eurasia we must note that in Japan there were rumblings below the surface that now began to elevate this part of the world in a prolonged orogeny. There is not much we can say about the far eastern landscape itself at this time, but it may have been not very unlike the topography there today.

Turning southwards across the great eastern mouth of Tethys, we would find Australia, well up in the southern latitudes but with western Australia adjacent to India at about 30 degrees south and Tasmania lying due south at about 63 degrees. The continent was almost completely emergent from the sea. A few lowland coal swamp regions existed in the east. But on the way from the northern latitudes our route would pass over a remarkable shallow shoal area of the ocean where in the clear waters countless shellfish, corals, sea urchins, sea lilies and many other forms of life abounded. Between the shallows there may have been deep channels. Farther to the south-east the Pacific mobile belt encroached upon New Zealand. Here geo-synclinal sediments were to accumulate throughout much of the Mesozoic era. At the southern end of the world, Antarctica was, like Australia, very largely above the waves with perhaps local lakes and swamps forming from time to time. Our knowledge of that continent in the Triassic is very limited indeed, but Antarctica seems unlikely to have reached its completely polar position.

By now the fact has been borne in upon us that, surrounded by its companion continents, Africa has had since Precambrian times a relatively tranquil geological history. Even during the Mesozoic era while it was swinging north towards Europe it was prone merely to small incursions of the sea around its margins. Little is known of what was taking place deep in the interior. Only in the Karroo Basin do many rocks remain to tell us what happened there in Triassic time. As we have seen, the Karroo Triassic rocks are justly famous for the enormous wealth of fossil reptiles that they contain. There must be hundreds of millions of animals represented by these bones and most of them were extraordinary reptiles that had many mammal-like characteristics. Herbivores and carnivores alike lived in a variety of habitats in the open country along the river banks and in the lakes and swamps. Much of the ground was rather barren and arid but at least in some regions and towards the end of the Triassic the climate was humid enough for swamp and peat deposits to form. As the Triassic period faded away there was a remarkable event quite unlike anything that had occurred here before. Basalts flooded out over the landscape from fissures and, less commonly, from volcanoes in the south-east. Flow upon flow erupted until a thickness of over 1,000 metres of basalt had buried the old Karroo landscape. The possibility that the emission of these basalts was connected with the break-up of Pangaea 2 and an accompanying fracturing of the rigid basement seems inescapable.

To the west of Africa South America was beginning to emerge as a separate continent splitting from its African counterpart. Like Africa, most of eastern South America stood well above sea level. Only in the Andean mountain belt are Triassic marine rocks much in evidence. Along the western side of the continent volcanoes and shallows prevailed, while inland deserts and arid country stretched in all directions. Locally there were small lakes and ephemeral rivers. The continent had little of the lush vegetation for which it is so famous today.

Allegro Moderato—Jurassic
The rifting and drifting movement begun in the Triassic period continued at much the same pace for the next 45 million years or so. The rifts extended and new branches of them appeared in several parts of the world. There was a fine new growth of ocean floor between North America and South America–Africa with an opening on the North Atlantic rift well beyond the region of the British Isles and perhaps into the Davis Strait between Canada and Greenland. A branch gulf had begun to open and edge north-western Spain away from Brittany. North America was continuing to drift to the north-west while Eurasia seems to have rotated clockwise to a small extent. This gentle movement of Eurasia had in fact the continuing effect of closing up Tethys and the crust was continually being resorbed in the Tethyan Trench.

The slow splitting along the Indo-Antarctic and Pan-Antarctic rifts

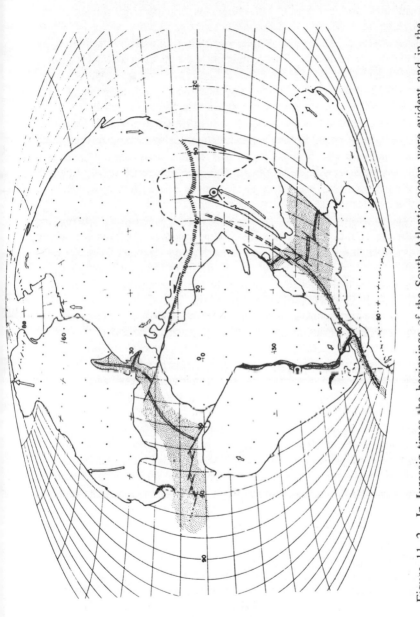

Figure 11-2 In Jurassic times the beginnings of the South Atlantic ocean were evident and in the Indian and North Atlantic oceans rifts were growing as new ocean floor (stippled) was produced. The arrows show the directions and relative amounts of drifting taking place. The black dot and circle mark a 'hot-spot' of volcanic activity.

(From Dietz and Holden, courtesy of R. S. Dietz.)

continued and perhaps at this stage the first deep fracturing of the African and South American blocks began. Vast areas of South America and south-west Africa were simultaneously inundated by great floods of basalt. There must have been spectacular eruptions, not violent or very explosive but a steady, relentless, pushing out of lava. The same kind of thing, you will recall, had happened over a shorter while in south-east Africa during the Triassic period.

India continued its remarkable drift northwards towards Laurasia, or perhaps it would be more correct to say it was being pushed from the south and also drawn northwards by the movement of crust into the Tethyan Trench. Australia moved on much more slowly towards the north-east and Antarctica turned on its axis a little to the west but otherwise remained near its original home.

In repeating the grand tour of Mesozoic continents during the Jurassic period, say a mere 60 million years or so later, we would have been presented with some conspicuous changes from the previous occasion. Not only had the positions of the continents changed and the oceans widened or diminished between them, but the landscapes and their occupants were different.

The western coastal area of North America in particular was engaging in a series of spectacular upheavals. The Cordilleran geosyncline was producing a succession of deeps close to the coast where at the same time there were plenty of volcanic fireworks. It seems likely that for part of the time there was a long, narrow island running parallel to the edge of the continent from Alaska to Mexico. To the east of this long barrier much of the western half of the continent was flooded by shallow sea. Much of the adjacent land was low-lying, swampy, given to floods. In the hot, probably humid climate, plants grew in great profusion. Amidst this steamy landscape and in the seas nearby lived some of the largest vertebrates of all time; the dinosaurs and the marine reptiles have left their bones in this region as relics of an exciting chapter in the history of life.

Living things apart, this was an area where remarkable geological events were taking shape. The Nevadan orogeny was now under way, culminating in the vigorous uplift of a long offshore island belt and the intrusion of huge, deep bathyliths of granite. Arms and gulfs of the sea, locally cut off by the rising land, dried up to leave crusts and layers of salts behind. Much of the great inland sea was drained and a monotonous complex of river sands and muds smothered much of its old site.

Around what is now the Gulf of Mexico and in Mexico the sea

floor dropped steeply away from the land. It was a development no doubt linked to the opening and deepening of the North Atlantic.

At this time, we may be fairly sure, the mid-Atlantic ridge was an active volcanic region, but no eruptive rocks are present in the nearest Jurassic formations of either eastern North America or westernmost Europe. In Europe itself Jurassic geography was changing rapidly, transgressions and regressions passed to and fro over a continent that managed to preserve much of its Palaeozoic geology as rounded uplands and highlands, despite the erosion of the Triassic period. The attack of the seas was twofold. It came from Tethys and from the Atlantic. The old Palaeozoic lands stood out as low islands, peninsulas and swells while the Jurassic seas ebbed and flowed around them. They were probably covered with a fairly persistent mantle of lush vegetation and the rivers that flowed from them brought mud rather than sand into the sea. In the clearer parts of the sea, corals and shellfish were abundant and great reefs and shell banks grew up not far from the coastlines. At the end of the Jurassic the southward retreat of the shorelines, which may be regarded as a response to the slow squeezing of the Tethyan geosyncline, served to leave most of Europe above sea level. Extending far to the east, Siberia formed a great eastern continent at this period with a few rather small inland drainage basins harbouring the sands and sediments washed in from neighbouring hills.

Australia presented a similar picture, with small embayments around the coast and a few large inland basins where steamy swamp conditions prevailed. Open lakes occupied parts of eastern Australia throughout much of the late Jurassic period. In New Zealand the Triassic geosyncline persisted on into the Jurassic until late in that period when it went the way of so many other geosynclines; the oceanic crust beneath it crept inexorably in from the east and the sediments on the sea floor were heaped up into mountain arcs.

New Zealand, of course, occupies part of the Circum-Pacific belt that was already undergoing the wrinkling and contortion of the crust, the plate destruction, that continues to this day. The rest of the western Pacific margin seems to have suffered during Jurassic time with ocean deeps forming north from Indonesia to the Siberian Arctic coast.

A glance at India is enough to reveal that only minor invasions of the sea around the west coast took place. For much of the subcontinent the Jurassic seems to have been a period when erosion of a land surface probably well covered by vegetation proceeded without a break. On the north coast of India the waters of Tethys may have advanced and retreated while the ocean floor was deformed and crammed into the Tethyan trench as the subcontinent drew northwards.

To Africa next; our inspection here shows a series of large embayments flooding south into the dusty Sahara area, but never for long. Elsewhere it was a scene of arid or tropical uplands, probably not very different from the landscape of today.

Southwards into Antarctica, our glance alights upon a Jurassic legacy of volcanic rocks and some sandstones remarkably full of plant remains. There were earth movements on a wide front before these rocks were formed, but the geography of the day is as yet obscure. The equator crossed the continent roughly from Cairo to Ghana. No Red Sea as yet separated Africa and Arabia, and South America still lay close to the west coast of Africa.

At the conclusion of our tour we return to South America where the Andean geosynclinal region was active along the west coast. South America was perched at the edge of its plate, the trench was close to the west. Sudden floods of the sea into parts of the interior of the continent came now and again from the west, but late in the Jurassic period there was a widespread uplift. As on most such occasions, it was a signal for vigorous volcanic uproar and the picture closely resembled that in North America.

Con Brio—Cretaceous

By late Mesozoic time the sundering of Gondwanaland became almost complete. In the Cretaceous period, as we have already noted from the evidence in Brazil and West Africa, these two blocks were now well separated. The Atlantic rift had spread perhaps as far as north Greenland at 60 degrees and as far south as 60 degrees before swinging round to the east and north-east as far as Madagascar. By now all the continents except Antarctica were moving northwards. Africa made as much as 10 degrees to the north. Africa-Arabia and India were moving towards the Tethyan Trench and the Tethys Ocean was narrowing rapidly. Part of the effect of squeezing and narrowing of the ocean floor was that the thick piles of sediment there were pushed up into arc-like ridges and islands, the scars that preceded the Himalayas. At the western end of Tethys Spain had slewed to the south with the opening of the Bay of Biscay.

Somewhere off the west coasts of North and South America the north-south trench system that must have existed throughout the

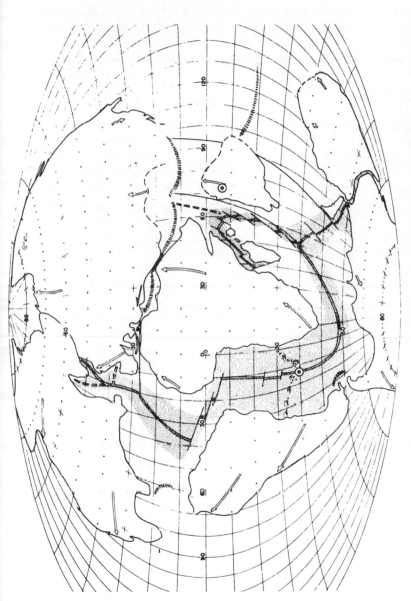

Figure 11–3   Some 65 million years ago, Cretaceous days saw the joining of the North and South Atlantic oceans and thus perhaps the spread of a more temperate climate. Huge areas of new ocean floor were produced and 'hot-spots' in the South Atlantic and India produced volcanic activity.

(From Dietz and Holden, courtesy of R. S. Dietz.)

Mesozoic era continued to extend its influence. It had long been important as the zone in which old crust was returned to the mantle below. Now both western continents had by Cretaceous days reached this trench and perhaps began to encroach upon it. The effects of this encroachment were vigorous upheavals in which the Mesozoic ocean sediments now were transformed and began to rise as the great Cordillera.

The super-continent of Laurasia had thus by the end of the Cretaceous almost ceased to exist. Only perhaps at the very northern end of the Atlantic was North America still in contact with Eurasia. As the continents separated so, it seems, were they to suffer what possibly were the most extensive transgressions to occur in Phanerozoic time. Geologists in North and South America, Africa, Europe and Australia have all been very impressed with the great white blanket of Cretaceous formations that spread out beyond the limits of the earlier Mesozoic marine incursions. Then there is the remarkable plankton-given character of the chalky limestones that make up this cover. It was as if the conditions of the mid-Ordovician epoch were being repeated. Much of the continents lay warm beneath a thin film of sea and lime-mud.

In Europe there was a steady spread of the waters northwards out of Tethys. Life in these waters was most abundant and in the lowlands there were steamy swamps and forests with an abundance of all manner of living things, some of them very strange indeed. Parts of the inland country, however, may have been arid with the hot winds blowing clouds of dust out to sea in the seasons of the year. It was much the same picture in North Africa, away south of the Tethyan sea. There the waves had by late Cretaceous days pounded their way into great embayments in the north coast, while in the west a link was made with the waters of the South Atlantic via the Gulf of Guinea and Nigeria. In fact all round the margins of this great continent the tides rose to cover a little more and a little more. In parts of East Africa freshwater deposits, sediments that buried the corpse of many a large dinosaur, were gathering in lagoons, lakes and river mouths.

All across the present Sahara area was a vast shoreline and sand area where the winds of the African interior blew north to the shores of Tethys.

East from Africa through Turkey, Iran and into the site of the great Himalayas today, Tethys continued almost uninterrupted. Here, too, however, and for a time at least, the sea may have temporarily been displaced by local uplifts. Then, too, in other regions there were deep submarine eruptions of basalt and water-laid volcanic ash. As far east as Assam and Bengal these volcanic episodes and uplift occurred, while in Burma there was a veritable centre of deep-seated volcanic intrusions and disturbances.

The great inland lakes that had occupied much of eastern Australia in the Jurassic persisted into Cretaceous times. Some of them were slowly filled in with swamp forest peats and bog deposits, long after to become coal seams. On occasion the sea broke into some of the lowest and most coastal lakes to produce gulfs and embayments, but in time these too were filled with sands and muds from the land round about. In the west the encroaching sea was more persistent and although it never got far inland it was never far away.

The geosynclinal or trench region that had established itself in the Indonesia–New Guinea area persisted. It continued to be restless, troubled with sinuous tracts of deep water and constantly changing shorelines in the Cretaceous period.

The Cretaceous scene in New Zealand was very different from that of the earlier Mesozoic. Although there was some geosynclinal activity in North Island, with volcanoes erupting, most of the region was continental upland, being eroded away, or a lowland area of sandy plains merging with the sea to the east. Away north in Japan the previous geosynclinal activity came to a climax with earth movements piling up the strata, with intrusions of igneous rocks and vulcanicity. When these had expended their not inconsiderable energies, sediments began to accumulate in the large hollows of the irregular land surface.

At the end of the last of these three periods of geological time, Triassic, Jurassic and Cretaceous, we have reached the end of an era in which the present shape and structure of the earth's crust were broadly set out. Large geosynclinal regions had taken form and had been compressed as the crustal plates grew and shrank and as the continents had set out on their latest routes. The great mountain ranges of the Cainozoic had their origins in these geosynclines. A glance at the map of the world shows how our modern mountain ranges are all situated where Tethys and the circum-Pacific geosynclines once existed.

Over much of the world the sea seemed to be retreating from the margins of the continents when the Mesozoic began. Shortly afterwards it regained the initiative and the Mesozoic era saw a succession of broad attacks upon the continents by the sea. Although the most effective of these, flooding more of the land than any since the

Devonian period, came towards the end of the Cretaceous period, the Mesozoic era closed with the continents apparently emerging from the waters once again.

When the seas were not in possession, the land was subdued and low-lying and with much of the natural drainage flowing away inland from coastal uplands. Of all the continents, Asia perhaps had most successfully withstood the creeping submergence.

We originally defined the Mesozoic era on the basis of the fossils its rocks contain. They seem to be sharply distinguished from those of the Palaeozoic and the Cainozoic, but this is more apparent than real. Despite the evidence used by the Catastrophists in Europe to demonstrate the suddenness of the beginning and end of Mesozoic time, we now have better evidence to show that there was no sudden great event to mark off this span of time. Where the fossil record is more complete it is very difficult to know which particular criteria to select as marking the first and last events of the Mesozoic. What is certain is that the Mesozoic was merely a long time in which the passage of events, especially continental drift, continuously modified the face of the earth from what it had been in the Palaeozoic to what it became in the Cainozoic. The evolution of climate and of animals and plants proceeded similarly without a break, and those are the topics of our next sections.

Halcyon Days—Mesozoic Climate

Cold though the end of the Palaeozoic era had been in the southern continents, and in the northernmost latitudes too, the Mesozoic seems to mark a warm spell. For most of the 160 million Mesozoic years much of the world experienced a rather equable and warm climate.

The positions, and to some extent the relief, of the Mesozoic continents are better known than those of the Palaeozoic. It is something of a temptation to draw on the Mesozoic globe patterns of ocean and wind circulation that conform largely to those of the present day. In the case of the Triassic geography this is not likely to be a very rewarding occupation, but the world map of the Jurassic and the Cretaceous periods does show the continents nearer to their present places on the globe.

Between the Triassic period and the end of the Cretaceous both polar regions were occupied by ocean rather than continents and this would tend to favour an equable world climate. The great Pacific Ocean extended westward into Tethys and its westward-sweeping equatorial currents may have penetrated far into Tethys, returning via two currents, one along each shoreline. Thus warm waters may have travelled north and south respectively along the south Asian and the northern Gondwana coastlines. Cool returning currents may have turned equator-wards from the high latitudes of the west coast of the Americas. Far inland there would have been little influence felt from all these currents. Hot, arid years would pass with the occasional or rare storm penetrating the interiors of the continents. The mountains pushed up at the end of the Palaeozoic and lining the northern side of Tethys tended to keep moist air out of the Eurasian interior, but hill storms and rains would occasionally wet these dry regions. In eastern North America the regime was much like the European.

In the Jurassic period the picture began to look more attractive, less desiccated. Sea level rose to flood large areas of the continental margins and the opening of the Atlantic Ocean allowed a narrow arm of water to reach into the continent. The Western Cordilleras were not yet high or broad enough to exclude moist air from the interior of the Americas. The polar regions were cool or temperate, and not at all cold as they are today. Tropical climates prevailed over much of the present temperate zones and even such large animals as dinosaurs were able to live as far north as the Arctic Circle.

Only towards the end of the Cretaceous period did the climatic pendulum seem to swing in the reverse direction. To what extent climatic change set off the train of extinctions at the close of the Mesozoic era is uncertain, but the connection between the two is important in the eyes of many geologists. With a slight drop in world average annual temperatures perhaps there was also more of a distinction between summer and winter. Seasonal changes became more pronounced. The cycads and about half the species of early flowering plants died out and the conifers began to extend their realm little by little from the cooler areas.

Another line of evidence from fossils which indirectly suggests that a cooling phase began late in the Cretaceous concerns the microscopic flora of the chalky limestones that are so common towards the top of the Mesozoic section. During the time when the shallow, sunlit, chalk seas were spread so extensively across the continents there was a remarkable boom of floating single-celled algal plants. For something like 30 million years they may have been amongst the most abundant plankton. Many of them secreted a box or case of minute limey platelets, known as coccoliths. It is these platelets that have built up the hundreds of metres of chalk. A rough estimate places these algae at about nine times as abundant in the late Cretaceous as they are today and in such abundance that their photosynthetic activity may have tilted the balance of the atmosphere in favour of oxygen and depleted it of carbon dioxide. As on previous occasions when that has happened, the 'greenhouse effect' has been diminished somewhat and world mean temperatures slowly dipped a little.

From the Mesozoic seas and from palaeontology we have another series of clues which may be used to verify our ideas about the climate. Here geochemistry comes to our aid again, and as in the case of the radiocarbon 14 method of determining the age of recent geological events, we owe the idea basically to American geochemist Harold Urey. While pondering about the relative abundance of isotopes in nature, he conceived the notion that the abundance of the oxygen isotopes 16, 17, and 18 in sea water depends upon the temperature of the water. If calcium carbonate is deposited from sea water it will contain the ratios of oxygen isotopes present for that temperature. Analysis of ancient shells should reveal the isotope ratios for the habitats of long ago. In other words, it should give the temperatures when the shell was formed. It took a few years to iron out the snags in the analytical techniques but the first practical demonstration was a great success. It was carried out on the skeleton of a squid-like fossil from the Jurassic rocks of Scotland. The concentric layers, like tree-rings, in this material seemed on analysis to indicate that the creature was born in the summer seas with a temperature about 20 degrees centigrade and four years later died in the spring when the temperature was only about 16 degrees centigrade.

This neat way of finding palaeo-temperatures, however, has to be treated somewhat carefully because we now know that not all the oceans have the same isotopic composition at any given temperature. Rain and river water, evaporation and depth are all factors liable to upset the ratios. For the moment we have to limit our attention only to the fossils of animals that lived in the open oceans. Even that precaution leaves us with the uncertainty that ancient oceans may have had different isotope ratios from those of today.

Although the details provided by this method may be somewhat suspect, the trends in the rise or decline of temperature over parts of the Mesozoic era seem to be shown clearly enough. The overall trend was downwards but there was a peak at about 80 million years ago and a decline thereafter. Many palaeontologists have seized this information to help account for the extinction of the dinosaurs.

All in all, the hazards and uncertainties that beset our attempts to understand past climates are still very formidable. With approaches ranging from geochemistry to mathematical models based on present climates, we are able to make more detailed suggestions than we can merely on the basis of fossils and the overall features of the sedimentary rocks. The Mesozoic offers us an era when in general the world climate seems to have been given only to slow and not very profound change and to have been equable rather than extreme in its seasons. It is easier to suggest what the Mesozoic climate was like than what prevailed in Palaeozoic times, but the task is more difficult than for the Cainozoic.

Mesozoic Life

The changes which occurred towards the end of the Permian period put paid to many of the long-established Palaeozoic organisms. As some kinds of animals and plants became extinct new ones took their places. Among the typically Palaeozoic groups to fade away at the end of that era were certain large protozoans or foraminifera, the trilobites, the strange segmented eurypterids, the rugose corals, many bryozoa, echinoderms and brachiopods. This was a fairly substantial portion of animal life in the seas and into the vacuum caused by its disappearance came many new vigorous kinds of creatures, eager to explore and colonise the available environments. The web of life was becoming more complicated, richer in the variety of its stands.

On land the end of the Palaeozoic era brought about a rather strange situation. Little change took place in the plant kingdom. The plants of the early Triassic are not unlike those that existed in the late Permian period, and suggest that there were tropical and temperate climatic zones. Conifers seem to have been common in the northernmost and southernmost lands, but towards the equator the flora was more varied. There was a bigger variety of plants in the warmer latitudes, just as there is today. Tropical cycads, the ginkos or maidenhair trees and ferns were the principal contributors to this scene, but no doubt there were were also many kinds of smaller plants that occurred in profusion. Mosses and other lowly growths abounded in the damp places, to be sure. In other areas, such as western Europe, the Triassic period produced deserts with true desert floras. Nowhere, yet, was there any sign of the grasses. They did not arrive until the Cretaceous period—and what a difference they must have made to the drier upland and plainland areas. When the grasses had achieved a hold upon the land new habitats for animals were available. There might well have been some kind of race amongst the land-dwelling creatures to be the first to secure a hold on these new territories.

However, most of the established four-legged inhabitants of Pangaea 2 were to become extinct and to be replaced by new, distinctive Mesozoic descendants. For the next 160 million years animal life on earth was to carry out some of its most spectacular and successful experiments.

The fact is, however, that many of these experiments were seemingly slow to get started. From the early Triassic rocks there are comparatively few fossils. Most of the early record of Mesozoic life has come from the strata formed in the mid and late parts of the period. As ever, marine fossils are the commonest. They were some of the first fossils to catch the eye of early man in Europe, and they occur in truly prodigious numbers in many rock formations. One cannot but be impressed by their numbers, variety and, in many cases, their beauty.

Life in the Mesozoic seas must have become prolific indeed. The clams evolved rapidly and came to rival, eventually to surpass, the brachiopods in their colonisation of the sea floor. Perhaps none were more successful than the oysters and several related groups. They spread into most seas and some of them became giants of their kind, surely the sign of successful adaptation to a habitat. One bivalve group produced an enormous thick cup- or coral-like shell, sticking up from the sea floor with a small cap-like shell to cover the animal. In other regions of the sea floor the corals flourished; these were probably the tropical areas where the seas were warm, clear and shallow and extended well beyond the present tropical latitudes. The Mesozoic corals built their skeletons of the lime mineral aragonite, rather than calcite as the Palaeozoic polyps did. The reefs they constructed were in most ways comparable to the teeming coral cities of the modern Indian and Pacific Oceans and the Caribbean. And in the same way the Mesozoic reefs offered shelter and food to many other kinds of sea creatures. Between barrier-like reefs and the coasts of Europe there were lagoons of relatively quiet water. Starfishes, sea urchins and crinoids lived on and around the reefs; clams, snails, brachiopods and, no doubt, countless creatures that never left fossil remains sought shelter and food in the recesses and hollows of the reefs. Crustaceans such as crabs and shrimps took the place of trilobites scuttling and swimming over the reef surface and beyond.

At sea the Mesozoic was the Age of Ammonites. Ammonites were members of the class Cephalopoda—the 'head-feet', the same group which was so successful in the Palaeozoic. Although nearly all the cephalopods became extinct at the end of the Permian, some forms resembling today's *Nautilus* did survive and some other rather similar but possibly more specialised coiled cephalopods—the ancestral ammonites—managed to stagger across the Mesozoic boundary. By late Triassic time they seemed to have recovered their vigour. They commonly grew to 10 centimetres or more in diameter and the patterns made by the suture to the inner septa became increasingly complicated. During the Jurassic period the shells of ammonites grew in some cases to five or six times this size. They were strengthened and corrugated by all manner of ribs, ridges and knobs.

There is no doubt that the great variety of Mesozoic ammonites reflects the fact that they became adapted to several different modes of life in different habitats. They seem to have penetrated into every corner of the shallow continental seas and inevitably they braved the oceans too, feeding on the plankton and smaller marine creatures. We can distinguish a sort of provinciality amongst them—Tethyan species and species of the Boreal Sea in the north, for example.

Late in the Cretaceous their numbers began to fall off drastically. By the end of the era the ammonites became extinct and only a few species of their hardy but possibly more primitive relatives, the nautiloids, survived. Their disappearance is rather a mystery. Perhaps they could not compete with the developing marine fishes—or were simply gobbled up by the larger fishes, reptiles and even the sea birds.

In the seas and fresh waters alike the Mesozoic fishes were more like their modern descendants than were Palaeozoic forms. The heavily armoured types of the Palaeozoic, and the spiny little acanthodians were all extinct. But the ray-finned and lobe-finned fishes and the sharks were much in evidence. The ray-fins included fishes of rather modern appearance but with heavy, thick, enamelled scales. They, too, became extinct ere the Mesozoic ended and their places were filled with schools of the active teleost fishes so common today. The sharks retained a firm hold on their carnivorous careers in the seas, but the lung-fishes and the lobe-fins declined in variety and abundance. Competition from teleosts on the one hand, and the aquatic reptiles on the other may have proved enough to send them packing from many environments they had previously enjoyed by themselves.

The amphibia might be said to have lost their ambition in the late Palaeozoic era, and in the Mesozoic they made no remarkable contribution to the evolution of the animal kingdom. The die for them had been cast and was to be modified only little between the Palaeozoic and the present when the frogs and toads hopped into the scene. Not so for the reptiles! The Mesozoic was truly their heyday. The lands, the seas, the oceans and the skies were all their domain. Among the many kinds of reptiles to evolve in the Mesozoic were lizards, snakes and crocodiles, turtles and, of course, the dinosaurs and the flying pterodactyls. Their great success story in the Mesozoic is that they spread into such a great range of different environments, evolved many splendid adaptations to these surroundings, produced offensive and defensive mechanisms and managed to stay abreast of geographical changes for about 140 million years. Only then did perhaps the pace of life, the rate of needed change and adaptation, exceed what they could manage. Reptiles diminished and mammals prevailed—the Mesozoic ended and the Cainozoic began.

The dinosaurs are so popular a group that we should spare them a little more space on these pages. The earliest of them appeared in late Triassic time and seem to have been quite small active types, bipedal, like chickens. Two main lines of evolution guide the development of the dinosaurs. One produced the giant herbivorous, quadrupedal *Brontosaurus* and its relatives, and also the bipedal *Tyrannosaurus*, at 6 metres high the largest carnivore ever. The other line includes both bipedal and quadrupedal giants, the armoured dinosaurs and the aquatic 'duck-billed' dinosaur. Every continent except Antarctica had its dinosaurs and in some parts of the world it seems these animals lived in herds or large groups. Some seem to have been destined for dry upland habitats, others for life in rivers, lakes or swamps. At least one species is regarded as a tree climber and another as a fast runner, speeding on hind legs away from danger when threatened. The great flesh-eaters were probably solitary or living together in small groups and in a wide variety of environments. Although we think of the *Brontosaurus* as living much submerged in water which helped to support his 50-ton bulk, the architecture of its skeleton suggests a land-dwelling creature. How it managed to consume enough food to sustain its vast bulk is not known. Perhaps a specially nutritious variety of plant-food was both plentiful and easy to obtain.

A basic enigma concerns the dinosaur eggs. It is always assumed that they all laid eggs, but only one dinosaur has been found indisputably associated with fossil eggs. The eggs of the largest dinosaurs would presumably be of a giant size and would have needed very strong shells to contain the fluid present in reptile eggs. To break out would have been a major problem to a baby dinosaur, unless the eggs had special shells or no shell at all.

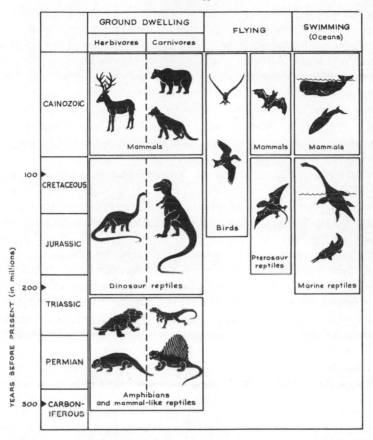

Figure 11-4 The Mesozoic era saw a vast expansion of the vertebrate animals with reptiles, birds, and mammals. The land, the waters and the air have seen a procession of animals adapted for life in those ecological realms. The pattern has remained rather similar despite the passage of scores of millions of years and the replacement of one group by another.

The dinosaurs and other reptiles were immensely varied in their adaptations to Mesozoic environments. We shall never know the full range of their virtuosity in this respect. It has only been exceeded by that of the mammals in the Cainozoic era.

Before we turn to the mammals, the flying reptiles are worth a moment's attention. The first flying reptile was probably a gliding form rather than a flapper of wings. A single Late Triassic specimen is known from New Jersey and other fossils probably representing different small gliding reptiles of the same period have been found in Britain. The more efficient pterodactyls or pterosaurs of the Jurassic period had wing membranes supported by the tremendously long fourth fingers. They had feeble hind limbs and, like bats, would have been helpless on the ground. Although the smallest were only the size of a sparrow, the largest had a wing span of over 7 metres and probably fed on fish from the Cretaceous seas. The end for the pterosaurs came when they had to compete with the feathered, more active birds, and by the end of the Cretaceous period it was all over.

The mammals were not in fact absolute strangers to the scene. They had been scurrying around in the undergrowth while the reptiles had their day. Fossils representing primitive mammals, small and rare, but mammals nonetheless, are known from Triassic, Jurassic and Cretaceous rocks. They were mostly shrew- or mouse-like beasts and they presumably had had reptilian ancestors in the Triassic or earlier periods. With the ability to produce young alive and to suckle the infants, and with improved nervous, circulatory and reproductive anatomy, the mammals were quick to seize the opportunity the demise of many of the reptiles provided. In particular, they had the ability to survive in colder climates.

Another group of highly accomplished creatures which we accept as an important part of the animal kingdom—the birds—first appeared in the Mesozoic. In mid-Jurassic rocks occur the very rare remains of what appears to have been the first bird, *Archaeopteryx*. It was about the size of a dove, had a long, reptile-like tail but with real feathers, not scales, and it possessed teeth in its beak. How well it flew is difficult to say, but no doubt the free claws that projected from its wings were used in clinging to the trees or other objects. *Archaeopteryx* may have looked like a lizard with feathers, but it had crossed the bird threshold. Thereafter came many other kinds of Mesozoic birds, large and small. Some became excellent swimmers, retaining feathers and other bird-like characteristics but losing their wings and relying on their feet for swimming.

Mesozoic animal life was on the whole spectacularly different from that of the Palaeozoic, but similar changes were not to be seen among the plants. Not until the mid part of the Cretaceous period did anything really remarkable occur in the plant kingdom. Then, comparatively suddenly, the flowering plants seem to have blossomed into abundance. Only one family of flowering plants is known from the earliest late Cretaceous, but by the end of that period at least 67 families existed. What a difference this tremendous development of flowering plants must have made to the landscape. We tend, perhaps, not to remember just how much of the green mantle of vegetation on the earth today consists of flowering plants, trees, herbs and grasses. Their growth and occupation of many different environments must have affected the living conditions not only of other plants but animals as well. Bees, butterflies and other insects no doubt, were presented with ecological opportunities that were too good to be ignored. The feeding habits of other creatures may well have been greatly influenced, and new types of shelter were afforded.

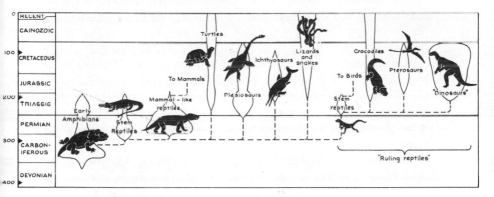

Figure 11–5　A riot of reptiles overtook the animal kingdom in the Mesozoic era and they established themselves in all continents and most environments. Only a few kinds survived into the Cainozoic era. The width of each column represents the relative numbers of the types of fossils known from each period. (After A. L. McAlester.)

The Second Armageddon

Towards the end of the Cretaceous period came another wave of extinctions. In perhaps a shorter spell of time than that at the end of the Palaeozoic era some of the seemingly secure and enduring lines of animal life came to an end. Their numbers diminished rapidly and finally they became extinct where previously they had been widespread and numerous. In the seas the ammonite cephalopods and their cousins the squid-like belemnites together with some families of bryozoa, echinoids and floating foraminifera all disappeared.

Among the vertebrates it was the great reptile clan that lost so many of its most formidable members. The dinosaurs, the flying reptiles and the great sea monsters were, within relatively few years, to disappear for ever. What brought about such decimation in the animal world and why were only certain groups struck down?

One might also ask if there were any features common to the extinctions that took place at the end of the Mesozoic and those at the end of the Palaeozoic. Really, there is very little that can be held to be significant on both occasions. The world was a different place, life had evolved from its Palaeozoic condition, yet there were perhaps two things that were in common. The seas were retreating from the edges of the continents and there was a small but perceptible climatic change. To what extent either of these was directly responsible for the extinction of any particular species will never be known for certain, but the climatic changes may have been in part due to a change in the 'greenhouse effect' brought about by plant life. The idea that cosmic rays flung out from the explosion of a star might have caused some forms of life to disappear from earth in the past was first produced by the Russian astronomer Shklovsky some years ago. Since then geologists and palaeontologists have alternately favoured it or rejected it as an explanation of what happened from time to time in the organic world.

Dale Russell, an authority on Canadian dinosaurs, has given much thought to the causes of their extinction. To him the idea of a colossal stellar explosion, a supernova, inflicting an enormous dosage of cosmic, gamma and X-rays upon the earth, seems to provide a possible source of trouble for the last of the dinosaurs. Such cosmic events are rare but not unknown within historic times. The chances are that once every 50 million years or so our solar system will come within 100 light years or less of a supernova. The radiation from this source would be strong enough to raise the dosage of gamma rays reaching the earth to a level adequate to affect adversely many forms of life. The largest animals in the shallowest waters and on land would have been affected most.

Russell also points out that the X-rays would be absorbed in the upper atmosphere, which would in the process become very hot. An event of this kind would in all probability cause turbulence, causing hurricanes to force masses of moisture laden air into the upper atmosphere. Once high above, the wet air would condense into a canopy of ice clouds, turning much of the sun's heat away from the earth. As this continues, the temperatures below decline, to the discomfiture of animals and plants adapted to tropical conditions. The chill lasted only long enough to upset the ecology of highly evolved terrestrial life, such as dinosaurs and flowering plants, and the largest animals at sea, but not long enough to produce ice caps. Ere long climatic conditions were much as they had been previously—not the same, but enough to allow modern tropical vegetation to burgeon and to favour the rapid emergence of the mammals as masters of the world.

CHAPTER XII

Cainozoic Climax

OUR last chapter is concerned with a mere 70 million years, a comparatively short span of geological time. Nevertheless, it is the era about which we can say most, the era in which the world has acquired most of its basic geography. Cainozoic rock formations are more widespread and less disturbed than those from the earlier geological eras. Our chances of making accurate reconstructions of Cainozoic geographies are thus that much better. The continents as we know them were distinct entities rather similar to their present shapes and growing more so as time went on.

The fossils from the Cainozoic rocks also have greater similarities to living organisms than have fossils from older formations. This makes it easier to imagine what the Cainozoic animals and plants looked like, how they lived and why eventually so many of them became extinct. The Cainozoic era has rightly been called the 'Age of Mammals' because mammals, including man, have been the most advanced forms of life on earth. Man likes to regard himself as the crowning evolutionary achievement, now largely in control of the rest of the biosphere, and his own environment, if not the world itself. There is certainly no doubt that he has become in a very short time one of the most important and effective factors in the present state of life on earth. His influence upon life in the future will probably increase greatly rather than diminish.

Man has also become a geological agent, a factor in the geological cycle. He speeds up erosion here by farming or other activity, delays it there by construction. He is mineral-hungry and always thirsty, in need of water and fuel. He has begun to wonder how long his planet can continue to operate its life-support systems to his benefit. Perhaps the end is in sight. Small comfort, then, to be told by the geologist that for the planet itself the end is by no means so obvious or near. The extinction of life on earth should, on the basis of the old classical criteria, mark the end of the Cainozoic era. 'Recent' life would have come to an end. It matters not whether the extinction be natural or man-made.

There do not seem to be any serious or conspicuous reasons why life should become extinct in the near geological future if only the human element can be kept in check. The geological future itself may well extend for another 10 thousand million years, by which time the heat of the sun will have been exhausted. Until then the course of geological events will probably conform much to the pattern that has evolved during the Phanerozoic history of the planet.

Viewed in the broadest way, the Cainozoic era is dominated by the great events that replaced Tethys with the Alpine–Himalayan mountain belt, the drift of the continents (through several degrees of arc and latitude) to their present places, and the uplift of the Circum-Pacific mountains. After the retreat of the seas from the continental platforms during the last stage of the Mesozoic era the margins of the continents were flooded once again. In the Eocene period this rise in sea level reached its maximum and since then the waters have subsided in a series of oscillations. By Pliocene time, 10 million years ago, the continents had assumed their present outlines but a new phenomenon began to affect the earth. The climate grew cooler. So cool, in fact, that in the last million or more years several continental glaciations have chilled much of the northern hemisphere and no small part of the south. We are living with the effects of this ice age and there is no definite guarantee that another glacial period may not be ahead of us.

Ocean Floors and Moving Plates

During this last era about one half of the ocean floor has been renewed from the mid-oceanic ridges. In total this amounts to about one-third of the entire surface of the earth. One-third renewed in 65 to 70 million years is a rate to be reckoned with; and if this has been the rate throughout most of earth's history, Earth has indeed been an active, dynamic planet, quite different from the place our grandfathers had in mind, completely renewing its oceanic crust every 150 million years or so. The permanence of ocean basins is a myth!

The North Atlantic Ocean continued to grow during this time as North America moved westward. The North Atlantic rift gradually penetrated farther and farther north into the polar basin, quite separating Greenland from Scandinavia. This was the end of Laurasia and the final split may have been marked by volcanic activity in several places. It is interesting to see what the effect of this break between North America and Eurasia would have been. The Eocene rocks and fossils of Spitsbergen, now at latitude 75 degrees north, tell us that the climate was warm or sub-tropical, with coal swamps covering hundreds of square kilometres of lowland. After the separation the colder waters of the polar basin would have mingled with the North Atlantic. It was an event that was to have far-reaching consequences. Slowly at first, but with increasing speed as the breach between the continents widened, the kind of climatic conditions we have today, with incessant cyclonic disturbance, might have been initiated. The flow of cool waters with their nutrient salts into the Atlantic may have had an interesting effect in promoting the growth of plankton and thus eventually affecting all sections of the ecological chain. The closed North Atlantic Ocean circulation was, by linking with the polar basin, changed to a more productive system for supporting a large and varied biota.

At this time, too, things were moving in the Caribbean area of the Atlantic. The country south of the Gulf of Mexico may have been swung into its present position from the south-east. The island arcs of the Caribbean developed as this movement proceeded and only at the last stage did the Isthmus of Panama connect North and South America. It is largely a volcanic causeway, the result of fierce andesitic volcanism, which is certainly not yet finished.

In the South Atlantic area the ocean continued to spread while South America moved westward and Africa slightly northward. South America came into contact with the Andean trench during this movement and seems to have bent and displaced it. Africa of course could not move very far to the north because Europe was in the way. The European end of Tethys was closed and Europe itself may have been shunted northwards a little. To allow for this the eastern part of Eurasia would have had to swing southwards. Such seems to have been the case, with the great Eurasian land mass turning clockwise around a point near Lake Baikal in Siberia.

The rifts separating Arabia from Africa and penetrating deep into eastern Africa appeared during the Cainozoic era and linked with the ocean ridge system of the Indian Ocean. South-eastwards this ridge, with its many transform faults, swung down west of India and south-west of Australia, then eastwards to detach Australia from Antarctica. The ocean-floor spreading that followed was very vigorous, carrying Australia and New Zealand northwards. New Zealand became detached from Australia during this drift.

Meanwhile India had been carried on northward from its Mesozoic position due east of Madagascar and Ethiopia. In mid Cainozoic it encountered the southern margin of Asia, and an open Tethys no longer

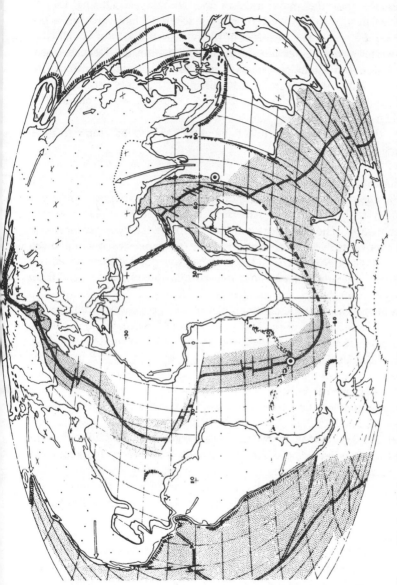

Figure 12-1 Cainozoic climax: the position of the continents today and the amount of sea-floor spreading that has taken place in the last 65 million years. The ocean ridge system has produced vast areas of new ocean floor (stippled) and the continents have been carried in the manner indicated by the arrows. There are still 'hot-spots' in the South Atlantic and Indian oceans.

(From Dietz and Holden, courtesy of R. S. Dietz.)

existed. The collision of India with Asia squeezed up the Tethyan sediments into the main arcs of the Himalayas. During India's passage northward its western margin seems to have crossed a hot spot on the crust. This resulted in the release of floods of basalt over the western part of the subcontinent. These were the Deccan traps. Even after India had passed over this hot spot basalt continued to erupt. In time, these eruptions, occurring as the crust moved north, built up a long basalt ridge on the ocean floor. It is now covered by thick growths of coral.

At the southern polar end of the earth, Antarctica seems to have swung slowly to the west.

Continents and Seas

While these movements were going on, the geosynclines along the western seaboard of the Americas and in Tethys were evolving dramatically. The shelf seas that spread from the geosynclines onto the continents were wide early in the Cainozoic era but narrowed later. Much of North Africa, for example, was flooded early in the era. Northern Europe suffered only intermittent incursions by seas from Tethys, reaching as far as the North Sea and Baltic areas at a maximum. The Atlantic continental shelf of Europe was submerged as far east as the flanks of the Massif Central of France, while in eastern Europe a shallow sea extended from the Ukraine, around the southern end of the Urals. From there one branch swung north to the Arctic along the eastern side of the Ural mountains, another continued eastwards towards India.

Large areas of the Atlantic coast of the U.S.A., the Gulf of Mexico and central Latin America were covered by continental seas, as were other coastal regions in Brazil, Peru, Argentina and Chile. For the most part, however, the American continents were emergent. The extensive orogenies of the late Mesozoic left a legacy of mountains, plateaux and intermontane basins. Many of these inland basins were the sites of great lakes and the accumulation of great spreads of sediment. Volcanoes erupted with vigour in the western parts of the Americas virtually throughout the Cainozoic.

For Africa the Cainozoic era has been one of steady but intermittent uplift and comparatively little other geological change. Over most of the interior erosion prevailed and on the northern, eastern and western coasts small gains were made by the sea. None of them lasted long and most of them took place early in the Cainozoic. However, the changes that did take place inland have given rise to some of

Africa's most impressive scenery. They are, of course, the changes associated with the development of the rift valleys and the highlands of East Africa. In a broad ridge-like uplift of the continental crust much of eastern Africa from the Zambezi to the Red Sea has been elevated to 2–3,000 metres above sea level and fractured along its crest. The famous rift valleys have resulted from the subsidence of long, narrow slivers of crust between the great faults. Volcanic outbreaks took place as the faulting continued. Mount Kenya and Mount Kilimanjaro are two of the best known of the splendid volcanic cones produced during these events, but there are many other less obviously volcanic mountains. The rift valleys are today the settings for great lakes as well as volcanoes, just as they have been since their beginnings. This region must on occasion have presented a truly impressive spectacle with its extensive lakes and marshes flanked by steep, even sheer, walls of rock hundreds of metres high, and with a background of tall volcanic cones, their plumes of smoke and vapour drifting across the blue African sky.

In Australia the Cainozoic was an era of comparative quiet and stability, although there were some hefty eruptions in eastern Australia during the Neogene. During late Mesozoic times there had been several lakes in the low plains of central Australia but by the mid Cainozoic they had all dried up.

Just north of Australia and in New Zealand it was a different matter. The Cainozoic there has involved a complex series of orogenies and patterns of deposition leading to the formation of shifting island arcs or ridges. Volcanic activity too has been extensive, aggressive and long continued. Throughout the arcs that stretch from the Aleutian Islands and Japan through the Philippines and Indonesia and into New Zealand, the Cainozoic has been a time of upheavals, eruptions, and the filling of deep sedimentary troughs and basins. It is as if the geological cycle had been turning at an extraordinarily rapid rate in this section of the globe. Volcanoes grew rapidly, only to be eroded away by the tropical rains and storms in short order. While on the eastern borders of the Pacific Ocean orogeny largely eliminated the sea from the American continents, along the western border of the ocean orogeny was in continuous conflict with geosynclinal deposition throughout Cainozoic times. Packed with incident, volcanic eruptions, earthquakes, the growth of deltas, the collapse of mountains, all took place on a spectacular scale, and yet Australia remained aloof and unmoved by it all. Her fossils show that little or no contact with the unruly lands to north and east was ever established.

K

Antarctic geology includes many Cainozoic sedimentary formations, volcanic rocks and several bathylithic intrusives in the west and
in the Transantarctic Mountains. Details of the geography are
obscure but the fossils present tell us that this continent basked in a
semi-tropical or temperate climate throughout much of Cainozoic
time. The great changes here came as early as the Miocene period
when snow and ice began to spread from the mountains eventually to
cover the entire land mass and nearby sea.

In common with those other stable continental areas, Africa, India
and Australia, Siberia had a rather uneventful Cainozoic history.
On its southern margin some of the most violent upheavals in the
world were taking place, but from the Pacific to the Urals and between
the Himalayas and the Arctic Ocean the land rose and sank in the
gentlest fashion. Cainozoic sediments were deposited only in faulted
basins between the highlands and at the eastern, northern and western
margins of the region.

A seaway linked the Arctic Ocean and Tethys east of the Urals
until Oligocene time when uplift and the closure of the Tethyan
geosyncline put an end to it. Siberia was from now on no longer
separated from Europe and when the climate began to cool the very
large land mass that was now Eurasia felt the extremes inherent in a
continental climate. It became an anticyclonic region, as cold in
Siberia perhaps as anywhere on earth in winter—a characteristic it
has retained ever since.

The inland areas of China and Mongolia, uplifted as Tethys was
destroyed, gave rise to rift or fault-controlled basins in which lakes
and swamps were spread. Here and there volcanoes erupted masses of
basaltic lava and ash as fault movements and local tilting took place.
Enormous areas of continental deposits extend over these regions.
Some of them have yielded remarkable fossil mammals.

Of all the great areas of Cainozoic rocks, that in North America
has been examined in greatest detail. Most of the petroleum wealth of
the U.S.A. derives from these strata so the work has had strong
support from industry. During this era most of North America was
actually above sea level and undergoing erosion. Only along parts of
the Gulf and Atlantic coasts and small areas of the Pacific coast are
Cainozoic marine sediments preserved. In the western plains and the
Cordillera, however, it is another story. The erosion there and the
accompanying volcanic activity have, between them, left a tremendous patchwork of volcanic and continental rocks from the Arctic
Ocean to Panama.

The Cainozoic sedimentary formations of the east coast thicken to seaward, spectacularly so in the Gulf of Mexico. They reach a thickness well over 10,000 metres in some regions, and together with the Mesozoic rocks along this eastern continental margin a total of 4,600,000 cubic metres of sediment may be involved. Deposition and subsidence have certainly been active here.

The Laramide orogeny in the late Cretaceous was largely responsible for the major features in the structure of the Western Cordillera. The spectacular relief of this region has largely developed as a result of later uplift and erosion, with additions from volcanic sources here and there. During this long period of uplift and erosion there were many wide lowland areas, some occupied by lakes, where sediment was caught. Thousands of metres of sediment were deposited in environments ranging from lakeside swamp to desert salt pan. Many of them are rich in the remains of extinct mammals, fish and plants.

Much of the north-west states of Washington, Idaho and Oregon was in Miocene time an upland area that had a topographic relief of 6,000 metres or so. At this stage it became the site of a flood of basalt lavas from many local fissures. The lava was very fluid and spread out rapidly, causing havoc to the local drainage. Here and there lakes were impounded, but not for long, as with the renewal of the eruptions they too were filled with basalt. By the time it was all over, some 1,500 metres of lava flows had accumulated, covering about 512,000 square kilometres.

Impressive as these events are, they compare with those taking place on the coast where the great coastal mountain ranges and the eastern Californian mountains were pushed up. Huge faults, great topographic relief and abundant signs of volcanic activity have become the legacy of these movements. The climax of the movements seems to have been reached in Pleistocene times and uplift is still going on. Fault movement still occurs, for as we have seen, the San Andreas fault in Southern California occurs at the junction of the American and Pacific plates. Volcanic outpourings have been prolific in the Californian region since Miocene time, and the Sierra Nevada has a blanket of lava and ashes about 1,000 metres thick.

As far north as Alaska and the arctic islands of Canada there are Cainozoic sedimentary rocks which were deposited under a temperate or even warm climate. Coal is a widely found Cainozoic rock in these regions. Volcanic formations too are common in Alaska with scores of active or only recently defunct volcanoes strung out along the Aleutian Islands and coastal mountains.

It might be said that for South America the orogenic crunch came in the late Cretaceous. At that time the giant bathyliths of the Andes were intruded and the whole region was raised. Little or nothing remained of the Andean geosyncline and during the ensuing era the sea only penetrated into the Andean area as narrow inlets and gulfs between the coastal ranges. Volcanoes built up tremendous piles of lava and ash, especially in southern Peru.

Within the Andean mountain ranges and on the land-girt eastern side of the Andes there were many places where debris eroded from the heights had settled. Truly prodigious thicknesses of silt, sand, gravel and boulders washed down by the torrents accumulated there and spread out from the foothills into the Amazon and Orinoco basins and onto the Pampas. From time to time faults underlying the Andean valleys moved, causing earthquakes no doubt, but also allowing some of the valley floors to subside and make room for more sweepings from the mountains to accumulate. These rift valleys were the sites of much volcanic rumblings and eruption, especially in Chile. There seem to have been two principal phases when Cainozoic earth movements further uplifted the Andes. The entire region from north to south was compressed in mid-Cainozoic time and it produced the usual backlash of volcanic eruptions, and a similar sequence of events took place just before the Pleistocene period began.

The Last of Tethys

The disappearance of one of the great oceans of the world and the growth of the grandest mountain belts known to us surely merits a section to itself. Tethys, as we have seen, was a major gulf, if that can be the right word, between Laurasia and Gondwanaland. It was an ocean separating continents, but an ocean that throughout the Meso-zoic and Cainozoic periods grew, narrowed, diminished, and van-ished. The Alpine-Himalayan orogenic belt that had its origins in the closure of Tethys and the squeezing of the ocean floor and shelf deposits into mountain areas, ranges and plateaux is one of the great mobile belts of the world today. Its relief, like a scar, hides an old wound in the crust of the earth, the long trench where the margins of adjacent crustal plates were in contact and where the plates from the south were drawn under the Eurasian plate.

There can be no doubt that the replacement of an open seaway by a tremendous array of mountains had a profound effect upon the circulation of the atmosphere. Asiatic weather in particular must have changed. The tropical and subtropical areas were reduced: temperate conditions became more widespread and extremes in climate north of the new ranges began to develop. All this was to have a significance for life and especially for the mammals.

Cainozoic Life

The changes taking place in the geography of the last 70-or-so million years in earth history must have been as dramatic as any experienced by the planet in so short a time. So, too, have been the developments in the evolution of life. Here, also, we may be more certain of our facts and we have many more facts about the life of this era than for any earlier times. We can make more detailed analyses of the floras and faunas of the Cainozoic seas and continents than for those of the Mesozoic or Palaeozoic. From studies such as these we may trace how geological changes and biological evolution may be linked. Not that this exercise can ever be regarded as complete; the best we can do is a good approximation to the truth.

The fossil record of the plants has shown us that, although the first flowering plants bloomed in the Triassic period, the modern flora began to sprout back in Cretaceous time. Nothing very startling within the plant kingdom seems to have occurred since then but a steady progress was made towards the development of the complex plant communities we know today in every climatic region of the world. And as the plant communities changed, so did the animals dependant upon them. With the spread of grasses across the great plainlands, new horizons were available for animals able to adapt to them, and no group of animals was more adept in this than were the mammals.

Following the extinction of so much of the plant plankton at the end of the Cretaceous, a new spectrum of these tiny plants appeared early in the Cainozoic. So rapidly did this take place that it is difficult to discover what effects it must have had on the diet of marine animals. The invertebrate animals of the Cainozoic seas seem to have found at an early stage the lines upon which they were to evolve. Cainozoic shells look very like modern ones; there have been no sudden spectacular changes or reverses. On the sea floor the clams and snails achieved a brilliant array of adaptation to the many kinds of environment that were appearing. So, too, did the tiny foraminifera. The sponges, corals, brachiopods and others seem to have achieved little fresh during the Cainozoic. Many modern kinds first appear in Cainozoic strata and of course there were local species that in their adaptation to special conditions, now gone, seem to have been bizarre, huge or freakish. The squid, octopus and *Nautilus* are the only Cainozoic representatives of the once great house of the cephalopods. The rise of the bony (teleost) fish which began in the Cretaceous period was swift and sure enough to present a serious challenge to the larger swimming invertebrates, and the cephalopods have continued to decline.

Arthropods, the walking and crawling invertebrates, have not been remarkable to any great degree in the Cainozoic, except for a continuing increase in the numbers of insects (and also ostracods). Insects have met the challenges of new Cainozoic environments with tremendous vitality and vigour. New modes of life, new mechanisms to meet the demands of hot, cold or other changing environments must have been acquired. The fossil record of insects is tantalisingly poor; its principals have not for the most part lived where their remains would be fossilized.

From time to time biologists and palaeontologists try to evaluate the rates at which organic evolution goes on. It is not an easy calculation, being dependant upon so many factors that are themselves not readily determined. Nevertheless, results of a kind are obtained. What is generally accepted is that life evolved in more complex and diverse ways in the Mesozoic than it had in the Palaeozoic, which was rather more than twice as long. In the mere 70 million years of the Cainozoic the acceleration in the rate at which the new species appeared on earth seems to have increased much further. Some groups became extinct during this time but there are more additions to, than subtractions from, the Cainozoic catalogue.

For the student of evolution, however, the most fascinating Cainozoic event must be the progress of the mammals. The great diversification of the mammals began very early in the Cainozoic. From humble origins in the early Mesozoic they remained in lowly circumstances until the end of Cretaceous times. Thereupon they hurried forward to enter every kind of environment in the ensuing era. In the sea, fresh waters and on land, they became the ruling vertebrates. Some even took to the air.

The Cainozoic stage, set with its continental connections linking Asia with North America across the Bering Strait, North with South America, and Eurasia with Africa in the Middle East, was ready for the fast action of the mammals. The forests, savannas and broad prairies that spread at the beginning of the Cainozoic would not have appeared very different from those familiar to us today. The dinosaurs had gone and the seas were deserted by the giant marine reptiles.

Smaller reptiles, it is true, remained in many environments, and of these the snakes, lizards, tortoises and crocodiles are the most familiar to us.

Inconspicuous for 100 million years or so of the dinosaurs' reign, the mammals had been evolving to the point where, once the dinosaurs became extinct, they could reach for the major role. As if to prepare for this, mammals had acquired several advantages. To keep up their high level of activity they had evolved a stable and relatively high body temperature, maintained by hair or fur, an active brain to guide their activities by learning and training, and improved methods of producing and feeding the young. Mammalian life is more complicated than reptilian life but it has enabled the mammals to penetrate and adapt to environments that daunt the scaly and cold-blooded.

Apart from the egg-laying mammals such as the rare duck-billed platypus of Australia, there are two basic types of mammal. The marsupials have pouches for carrying the young which are comparatively undeveloped at birth. The placentals retain the unborn young longer within the mother's body where the placenta forms a supply line for food from her bloodstream. Perhaps it was quite early in the Mesozoic that these three divisions of the mammals first evolved, and they have evolved independently ever since.

The earliest Cainozoic mammals were small herbivores and insectivores, not unlike their Mesozoic ancestors. Emerging from the undergrowth where for so long they had evaded the dinosaurs, they soon evolved larger forms in the new open habitats. In most continents marsupials and placentals strove to enter all the available ecological niches. The growing isolation of Australia seems to have favoured the marsupials and they produced their marsupial counterparts to many placental mammals elsewhere. South America too had its marsupial hordes, as we have already noted in Chapter VIII.

The first major action on the part of the mammals in the Cainozoic was to multiply into what have been called the archaic faunas. These groups largely failed to survive into later times. Many of the archaic mammals in North America and the Old World achieved a remarkable size, becoming five-toed browsers with bizarre horned heads. These titanotheres, or 'giant beasts', resembled modern rhinoceroses in appearance but were up to 2 metres high at the shoulder and 5 metres long. On their heads were massive bony horns. Large and small carnivores, some possessing hoofs rather than paws, were there too, no doubt keeping the numbers of herbivores down to a stable level.

Never slow to lose an opportunity, mammals had their representatives in the seas to take over, one might say, the role that the larger marine reptiles had just relinquished. Even as early as the Eocene period there were several kinds of whales, including a slender fearsomely toothed beast (*Zeuglodon*) as much as 20 metres long.

Then in late Eocene and early Oligocene times the archaic mammals were largely replaced by the ancestors of our modern mammals. All the present major groups were established, the hoofed mammals, the carnivores, the rodents, the primates all appearing on the scene within a very short period. The giraffes, pigs, deer and camels, the cattle, horses and rhinos all have recognisable ancestors in these faunas. If there is any one moment in the Cainozoic when the mammals could be said to have reached their zenith, it would probably be in the Miocene period, some 25 million years ago. It was the time of the greatest of the Alpine–Himalayan upheavals and with them came volcanic activity in many parts of the world. Temperate climates began to replace the tropical or subtropical and the plainlands were populated by a greater variety of mammals than ever before or after. From hereon until the Pleistocene ice ages the mammals steadily evolved, spreading out across all the continents except Australia and Antarctica in waves of migration.

Then came the changes of climate and geography over the last one or two million years that were so instrumental in guiding the emergence of our contemporary faunas. As the ice advanced and later withdrew from so much of the northern hemisphere the world assumed its modern aspects and many of the Pleistocene mammals faded into extinction. We are left with the remnants of the rich Pleistocene faunas. Our own ancestors, making their debut as hunters and exterminators, witnessed a procession of animals that has been diminishing ever since. Today there are about 5,000 known living species of mammals. Some of them, including the largest-ever mammals, the whales, are on the danger list for extinction.

If the Devonian can be called the 'Age of Fishes', the Cainozoic might be called the 'Age of Bony Fishes'. True, these versatile creatures had first made their appearance in the Devonian period, but until late in the Cretaceous period they seemed not to progress very far. Then began a quite remarkable spate of evolution. Coincident with the demise of the ammonites at sea the bony fishes became more and more prominent in the marine faunas. They produced feeders and foragers of all kinds and shapes, inhabiting every part of the oceans. Today there are over 5,000 known species of bony fish. Most live in the sea but there are plenty of species in brackish and fresh waters, at all latitudes. In the seas the bony fish more than hold their own against the ancient house of the sharks and rays, the cartilaginous skeletons of which distinguish them from the bony newcomers.

Figure 12-2 The extent of the Pleistocene ice caps in the northern hemisphere is shown by the lightly stippled areas, together with the directions from which the ice flowed. The Arctic and much of the North Atlantic Oceans were permanently covered by pack ice and icebergs may have floated as far south as Spain and the Carolinas.

Ice Age

At the end of the Cainozoic era we come to the Pleistocene and Recent or Holocene periods in earth history, a mere 2 million years or so. But for us they must be a very special 2 million years. In the first place, they witnessed the arrival of man, and there are few items in the geologist's scrapbook that attract more interest than evidence of the origins and antiquity of man. In the second place but no less significantly for the world at large, the Pleistocene covered what is usually referred to as the Great Ice Age. And the results of geological events in the last two million years are so fresh upon the face of the earth that they can be deciphered relatively easily. The story they reveal is not a simple one and scholars of the Pleistocene have found as much to debate as any other geologists.

Truly widespread glaciations and world-wide lowering of temperatures are rare events in the history of the earth. They occurred at least once, and possibly twice in the Precambrian, in the late Ordovician, in the Permo-Carboniferous and in the Pleistocene. It was, incidentally, only twenty years after the famous Swiss biologist Louis Agassiz became in the mid-nineteenth century the champion of the theory of continental glaciation that the Permo-Carboniferous glacial deposits of India were recognized. The knowledge of glaciation and its deposits that the Pleistocene and Recent periods have given us is of paramount importance in understanding the relics of ice ages long past.

However, the comparatively sudden cooling of the earth's climate does not provide good geological data for recognising or defining the beginning of the Pleistocene period. Today's polar regions and mountain tops have a glacial climate but no one would suggest that crossing the Arctic Circle or the 1,000 metres contour involves a journey back a million years in time. There is still some argument about exactly where and how to define the beginning of Pleistocene time. The Pleistocene–Holocene date line, i.e. the 'end' of the glacial epoch, is perhaps best marked at the end of the last rapid rise in sea level between 6,000 and 8,000 years ago.

In the course of the great growth and spread of glaciers, snow and ice to the extent of about 43 million cubic kilometres was dumped on about a third of the land surface of the world. That represents a lot of water, and the removal of this water from the oceans had effects in virtually every corner of the globe. Its influence extended to sea level, climate, topography and wild life. The ice itself had tremendous effects in the erosion of the land and the creation of deposits of weathered material and in causing the actual depression of the land beneath its great weight.

This last-mentioned feature is still changing the geography of once-glaciated areas. Ice seems to reach a thickness of about 3,000 metres in ice caps today, at which point wastage keeps pace with supply. Under a load of ice this thick, the crust of the earth is slowly but surely depressed. When the ice melts the crust returns gradually to its previous position, like a soft rubber ball. This is good evidence that the continental crust is capable of some plastic flow, and the rebound is shown most dramatically in parts of the Baltic, the Arctic and the

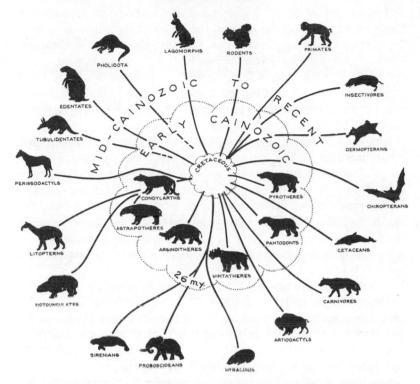

Figure 12-3   An explosion of mammals has taken place in the
Cainozoic era, with the rapid evolution of many different kinds
adapted to the varied environments of the post-dinosaur world. This
diagram shows only the orders of placental mammals that have
appeared in Cainozoic times. They far out-number the marsupials and
the other early and largely extinct mammals. (After a diagram by Lois
M. Darling in Colbert.)

Great Lakes regions of North America where Pleistocene beaches and
coastal features are now raised high above sea level and some are
tilted. The process seems to have been going on for the last 10,000
years and is still continuing. As the ice melts from Greenland so that
country, too, will rise from its present position, and Antarctica would
presumably behave in the same way.

    At their greatest extent the glaciers of North America covered the
land north of the Canadian border in the west, and north from a line
from Kansas to New York. In Europe the land north of the Bristol

Channel eastwards to the Ukraine and much of northern Russia and Siberia were under ice. So were all the higher mountain areas in both the north and south hemispheres and Antarctica was smothered. While it was snowing in the higher latitudes, it was raining in the lower. As a result there was a wholesale migration of climatic and vegetational belts towards the equator and in many regions large lakes were formed. There were giant inland lakes, virtual seas, in North America, Africa and Asia.

The glaciation itself produced changes to drainage patterns as the ice or moraine plugged stream courses, diverted channels and impounded new lakes. The ground around the ice sheets was frozen to great depths and only the top few centimetres thawed in the summer. This permafrost has had many strange effects upon the soils of the higher latitudes. Elsewhere the winds blowing out from the high pressure regions over the ice picked up dust (loess) from the glacial sediments and eventually deposited it in blanket-like masses and mounds beyond the ice-covered lands.

While all these features are local in origin and extent, the growth and waning of the ice sheets had also a world-wide effect on sea level. Today we are discovering submerged beach ridges and deposits far out on the continental shelf as deep as 140 metres, and there are fossils from beneath the sea that indicate that these regions were once above the waves. Sea level surely stood that much lower than at present.

Sea level appears to have dropped at least four times in the last 40,000 years. Locally there have been other oscillations superimposed on these. Lower sea level means more land exposed to a denudation by rain and wind, steeper gradients down which the rivers must flow. As the ice melted towards the end of each glaciation the swollen rivers rushed down their courses carrying heavy loads of sediment out across the continental shelves to be deposited in the seas and oceans beyond. These torrents may have cut the submarine canyons described in Chapter II.

In Europe and North America the interpretation of the sheets, spreads and dollops of clay, sand, gravel, peat and soil littering the surface of the solid rock formations has been a long and arduous business. For nearly 120 years geologists and others worked away at it, employing the same general principles as were used by the students of the older formations. Then halfway through this century the use of carbon-14 isotopic dating began to make an impact upon Pleistocene geology. It became possible to compare in detail the effects of glacial conditions on the continents and in the oceans.

To cut a long story short, the genesis of the Pleistocene deposits can now be related to four glacial periods with interglacial intervals. Since the last glacial phase, an interglacial has been in effect, beginning about 7,000 years ago. The ice first seems to have affected a wide spread about two million years ago. These icy phases and succeeding interglacials followed in the next million years or so and the last great advance began about 70,000 years ago.

The glacial periods have been exceptional times in earth's history and between them the climate has been milder and more uniform. Polar regions always were cooler, but not to the extent they are at present. From late Mesozoic time onwards to the Pleistocene, so the oxygen isotope analyses tell us, the world climate grew progressively cooler. The grand culmination of this was not a single great freeze-up but an oscillation of glacial advance and retreat. Glacial and inter-glacial phases followed one another in rhythms that each seems to have occupied between 30,000 and 50,000 years. What can account for both the long-term sag in world temperatures and these late oscillations? Are the two in fact related?

Despite the attention of generations of meteorologists, geologists and physicists, no-one can be certain that the exact cause has been found. In all probability continental glaciation comes about from the rare combination of several factors. Such factors might be changes in the radiation of the sun, changes in the motion or path of the earth, changes in the composition or transparency of the atmosphere and changes in the exchange and dispersal of heat from one region of the earth to another. There might seem to be plenty to choose from in this list of possibilities, but each is debatable. As prime or single factors each one may be dismissed, but in combination several of them could bring about a glacial interval.

As far as solar activity is concerned, there is simply not enough known about its variations to allow us to regard with confidence any suggestion that the sun's energy may be 'turned down' from time to time. The first convincing calculations to suggest that the irregularities of the earth's movement could cause climatic cycles were made by a Yugoslavian meteorologist, M. Milankovich. His data and ideas have been incorporated into a theory that the progressive dip of the mercury reaches its lowest every 40,000 years. This looks reasonable as far as the Pleistocene is concerned, but there were previous long periods when such a dip had nothing like the results that it produced in this epoch. The Milankovich effect, as this dip is known, has perhaps been ineffective, except when some of the other factors are also in play.

One such factor might be the degree to which solar energy is reflected from the atmosphere, rather than being absorbed. Clouds, dust and aerosols all can lower this ability to absorb the sun's heat. Volcanic dust can, we know, reduce this ability for short spells. The proportion of carbon dioxide may control the 'greenhouse effect' mentioned several times already on these pages. Ozone filters out the solar ultra-violet rays. What possible agency would have removed sufficient carbon dioxide from the atmosphere to set off the glacial phases of the Pleistocene is not known.

On the other hand, the general draining of the continents and the growth of extensive high mountain systems during the later part of the Cainozoic have been events that would have so changed the heat-reflecting character of the earth's surface that a general cooling resulted. Two American scientists, C. Emiliani and G. Geiss, have agreed that by the end of the Cainozoic the reflecting ability of the earth's surface was increased to the point where the general temperature was lowered enough for the Milankovich effect to come into play.

Two other Americans, the geophysicists, M. Ewing and W. L. Donn, believe that geographical factors in the Pleistocene may have given the final tilt to the glacial scales. Their ideas were put forward when continental drift and the apparent movement of the poles was first being confirmed by palaeomagnetic studies, which is an indication of their faith in the value of the work on ancient magnetism. They suggest that while the North Pole lay in the north Pacific Ocean a mild climate occurred. When, however, the pole became established in the Arctic Ocean basin late in the Mesozoic era, temperatures declined. The almost enclosed nature of the basin severely interrupts the heat exchange system of ocean currents which could carry from equatorial to polar zones heat received from the sun. The South Pole meanwhile, has been even more securely isolated from warm currents from the lower latitudes. It rests in the midst of a high and very extensive continent.

Ewing and Donn give us a picture of continental glaciers beginning to form first in the Antarctic, perhaps even in Miocene times. Evaporation from the Arctic Ocean transferred moisture to the local continents and glaciers began to appear in the upland regions there too. Eventually the Arctic Ocean itself froze over. The effect was to cut off the supply of moisture nourishing the ice caps and in a little while the rate of the melting process began to overtake the rate of growth of the ice. The glaciers went into retreat. The climate began to warm. The Arctic Ocean opened again, sea level rose, and precipitation could begin again.

A balance seems to be suggested with the climate swinging from glacial to non-glacial, again and again until either the polar position or the Arctic Ocean is changed in some way. Russian scientists have proposed to dam the Bering Straits in the calculated hope that this might ameliorate the climate. Their country would certainly benefit from an improvement in the climate, but if sea level were to rise other nations might not feel so happy about it.

The centres of Pleistocene ice accumulation lay well south of the Arctic Ocean and there is no consensus of opinion that evaporation from that ocean could provide all the moisture needed.

So to each theory there seems to be some initial objection, but there seems to be room to believe that by the coincidence of some of the possible factors we have mentioned, the Pleistocene ice ages were produced. We may be near the answer without recognising it outright.

Things To Come

From all that has appeared on the pages of this book it is obvious that the earth is indeed a complex system. The materials in its thin outermost layers are constantly undergoing rearrangement and change. Atmosphere and biosphere play a part in the geological cycle and all have been evolving for a long time. The evolution of the earth has affected the evolution of the biosphere, life itself. Presumably it will go on doing so for a long while yet.

What lies in the future is in detail uncertain, but the pattern of really major events is indicated by what we know from the skies around us. The evolution of man, in itself part of the evolution of the biosphere, has been influenced by the evolution of the earth, the solar system, the galaxy and ultimately the universe itself. Modern astronomy has detected suns older than ours, suns that are dying, stars that are dead. The process of dying in the celestial sense is recognised and to some extent understood. All scientists realise that the expenditure of energy in the universe can only proceed for a finite period of time, not for ever and ever. Some scientists believe that the end is already in sight. They detect in far distant space the regions where energy has run out, where matter is reduced to an infinitely small compaction and ceases to 'exist'. In due course such events must spread throughout the universe as we know it. It would be a terrifying prospect were it not so remote and were there not more immediate and serious dangers.

The earth will continue to evolve, perhaps for as long again as it already has. During this time there will be changes as profound as those which have occupied the first half of its existence and with effects equally important to life here.

Three aspects of this view of things to come concern us in these pages—the immediate and topical evolution of the climate, the longer-term changes to sea and land, and the ultimate fate of the earth and solar system.

Even within living memory the climate of various parts of the world has changed. In the course of human history it has altered and affected man's activities. The Norse explorations of the Atlantic in the dark ages—which were in fact climatically rather better than times before or since—were possible in fair summers and mild winters. The Norse colonisation of Greenland in the twelfth century was brought to a halt at least in part by the deterioration of the climate.

Man is beginning to understand the factors that control climate. He has at last within his grasp enough data to begin to construct mathematical models of the climate and how it behaves. From these we may extrapolate back in time to learn how the climate was in the past and project forward to predict what it may have in store for us in the future.

A return of snow and ice will pose severe problems to man and nature alike and it will demand much expense and ingenuity to keep the northern cities and countries from smothering. On the other hand, really warm spells with the ice caps of Greenland and Antarctica melting away will certainly raise the sea level by an estimated 70–90 metres. About 15 per cent of the present land area would be inundated. With flooded coastlands and most of the major cities and ports of the world then in jeopardy, the outlook for mankind, in its ever increasing millions, could be somewhat worrying, to put it mildly.

But perhaps there will be an increase of the 'greenhouse effect' that has been mentioned before. Such an event has been thought possible if the proportion of carbon dioxide in the atmosphere continues to rise at its present rate for another few thousand years. A more effective contributing factor leading to a new ice age may be the increase of pollutant aerosols in the upper atmosphere. Aerosols are collections of very tiny particles in the air. Smog is an aerosol. Aerosols could reflect the sun's heat and thus cause the earth to cool. American scientists at Goddard Space Flight Center think that another fifty years of aerosol production could cause another ice age to begin.

Further into the future we may probe to see how seafloor spreading and continental drift will affect our world. Presuming the present movements of the plates are continued, there will be some interesting developments. If new forces or patterns of plate construction and movement appear, the prospect is less predictable. It seems reasonably safe to predict, however, that the ocean ridges will continue to be active for a good many million years yet and so will the Pacific 'Ring of Fire'. The Atlantic Ocean will continue to widen and North America may tend to move northwards a little.

In the Caribbean there will be continuing volcanic activity and perhaps new island arcs. Mexico may suffer increasing paroxysms of volcanic activity and the movement of western California north-westwards west of the San Andreas fault may produce some interesting new features in that State.

In the Old World the great mountain ranges will decay and decline, while the American Cordilleras may continue to be rejuvenated by the activity at the plate margins beneath them. The ocean ridge rift system will begin to split Africa and Arabia along the lines of the present rift valleys and perhaps to extend north-westwards from Kenya to Libya, the central Mediterranean and even into central Europe. East Africa will drift away to the east or north-east of the main continental mass.

Australia will remain relatively unaffected by these events but in the Indonesia–New Zealand belts new lands may appear adjacent to the ocean trenches. Island arcs may come and go in the western Pacific. Antarctica remains something of an enigma: perhaps it will rest content at the southern extremity of the world, involved in little but a slow gyration or turning westwards as time goes on.

These changes may alter the circulations of atmosphere and oceans sufficiently to change world climate. Fragmentation of continents and a more intricate system of ocean circulation will make the world climate more equable, and possibly warmer.

The spread of warmer conditions into the higher latitudes, at least for a greater part of the year if not throughout, will mean changes in the biosphere. Periods of continental fragmentation in the past have been periods of evolutionary diversification among animals and plants. This could happen again, if man has not by that time reduced the biosphere virtually to a life-support system for himself alone. Evolution of the human species in the future is something that we had best perhaps omit from these pages.

In due course the pattern of ocean plate formation, movement and destruction will change, just as it may have changed in the past, and the process will slow down. Volcanism, orogeny and continental drift will not go on for ever, and in the end they will cease. Perhaps then, another 5,000 million years hence, other influences will press in upon the earth. Threats from outer space or at least from the moon and sun will be in the offing.

During the course of its existence the moon has exercised a kind of braking effect on the earth's rotation by raising tides. Once they were much bigger than now, because the moon was so much nearer to earth. In the future it will still be slowly inching away from us in bigger and bigger orbit. The radius of the orbit increases by about 1 metre per century. Tides will diminish and the spin of the earth will gradually slow down as the angular momentum is transferred to the moon. The day will grow longer. Eventually, however, the moon will edge so far out in orbit that its influence will be lost and earth will begin to speed up again. Earth's own accelerated rotation may at this point cause the moon slowly to spiral back. Having come close in, the tidal strains on the moon may fracture it to destruction and it will perhaps then circle the earth as a ring of asteroid-like fragments, but in a flat belt like the rings round Saturn. Chunks would almost certainly rain down on earth with consequences that can easily be imagined. This will, however, not be until about 5,000 million years in the future. We can all breath again.

Meanwhile, back at the centre of the solar system our sun will have consumed enough of its nuclear fuel to embark upon that stage of stellar evolution when it begins to puff up into an enormous size, and to turn blood red. It will become a red giant and fill up space as far out as the orbit of Mercury. What will have happened to Mercury can be left to the imagination, for by that time the temperature on earth, so much further out from the sun, will be in the hundreds of degrees. The oceans will have boiled away into an eerie shroud of steam.

This hellish state of affairs may last a thousand million years until the sun has disgorged its energy and begins to shrink rapidly into the condition typified by the stars known as white dwarfs. Its life will be over, the sun will be dying. What remains of frazzled and steamy earth will cool into an icy ball, its waters frozen into permanent ice. Eventually the light of the sun will be extinguished and at that time, several hundred thousand million years in the future, it will collapse under its own gravity into a dense mass, dead, inert, and actually smaller in size than the earth.

The planets will be at that stage still locked in orbit about the sun, dead and frozen. The satellites about the planets will begin to edge their orbits out into space. The heavens will have changed, the patterns of stars will be different, but no man will be here on earth to chart the skies.

Life on earth will have ended long before this. Be it through 'natural causes' or man's interference, life will have ceased long before this distant point in time. Earth, the spacecraft that has travelled so far and so long, will perhaps for a short period in her middle age have carried a cargo, scarcely a crew, of living things. She will have travelled from the unknown to the unknown, and *en route* for a brief span her voyage will have held the attention of one species of her inhabitants, and he may have got the story quite wrong.

Further Reading

Restless Earth, Nigel Calder, B.B.C., 1972.
Our Planet Earth from the Beginning, J. Sadil, Paul Hamlyn, 1968.
Continental Drift, D. H. Tarling and M. P. Tarling, Bell, 1971.
Physics of the Earth, T. F. Gaskell, Thames & Hudson, 1970.
Planets and Life, P. H. Sneath, Thames & Hudson, 1970.
Before the Deluge, H. Wendt, Gollancz, 1968; Paladin, 1970.
The Procession of Life, A. S. Romer, Weidenfeld & Nicolson, 1968.
A Guide to Invertebrate Fossils, F. A. Middlemiss, Hutchinson, 1968.
Prehistoric Animals, J. Augusta & Z. Burian, Spring Books, 1954.

Glossary

Absolute Age—Geological time in years. Compare with relative age.

Algae—A group of simple plants containing chlorophyll and capable of photosynthesis. Some are capable of secreting calcium carbonate and are important rock-builders.

Amino Acids—Nitrogenous hydrocarbons which serve as the building units of the proteins and are thus essential components of living matter.

Andesite—A volcanic rock intermediate in chemical composition between granite and basalt: typical of continental volcanic activity in mountain regions.

Arthropoda—A phylum of invertebrate animals with jointed legs. Spiders, insects, crabs and the extinct trilobites belong to this very successful phylum.

Ash—Small rock fragments and particles, commonly less than 4 mm in diameter, erupted from a volcano by explosive activity.

Asteroid (Minor Planet)—A small celestial body in the solar system normally having an orbit between those of Mars and Jupiter. The largest (Ceres) has a diameter of 700 km. Only 10 per cent of the thousands of known asteroids exceed 30 km diameter.

Asthenosphere—A zone of the earth between 50 and 250 kms below the surface where earthquake shock waves travel at much reduced speeds; possibly having less rigidity than other parts of the interior, it may be a zone where movement such as convective flow takes place.

Atom—The smallest divisible unit retaining the characteristics of a specific element. Atoms are defined by the atomic number and mass. The chemical properties of an atom depend on the atomic number.

Basalt—A fine-grained, dark-coloured igneous rock formed by solidification from a molten or partially molten state.

Bathylith—A very large mass of intrusive rock, generally composed of granite or granite-like rocks.

Brachiopoda—A group of shelled marine invertebrates now comparatively rare but abundant in Palaeozoic times.

Catastrophism—The belief that world-wide events of a different kind of magnitude from those now observed are needed to explain much of the origin and development of the earth and life.

Cephalopoda—Members of the mollusc phylum with relatively well-developed sensory systems, and with tentacles projecting from the head region. Many have shells, coiled or straight, and they are numerous as fossils in Mesozoic rocks.

Chert—A highly siliceous cryptocrystalline rock composed dominantly of opaline silica and quartz.

Conglomerate—A sedimentary rock containing rounded pebbles: equivalent to a consolidated gravel.

Continental Shelf—Shallow, gradually sloping rock surface from the sea margin to a depth of about 200 m beyond which there is a steeper descent, the continental slope.

Continental Slope—The submarine slope extending from the continental shelf to the ocean floor: relatively steep compared to the shelf.

Core—The central part of the earth lying below the mantle.

Craton—A long-stable region, commonly with Precambrian rocks at the surface or only thinly covered.

Crust—The thin outer shell of the earth separated by the Mohorovičić discontinuity from the mantle, 6 km to 70 km below the surface.

Cryptozoic—Referring to geological time before the beginning of the Cambrian period, about 600 million years ago; also known as Precambrian or Prepalaeozoic. Compare Phanerozoic.

Cyclothem—A vertical succession of sedimentary rocks which occurs many times in a formation and which denotes the repetition of physiographic and sedimentary events in a constant order.

Density—The ratio of the mass (weight) to the volume of a substance. This is expressed as Kg/cubic metre.

Dinosaurs—An extinct group of reptiles, generally large, confined to the Mesozoic era.

Dyke—A near-vertical sheet of igneous rock, or one which breaks across stratification. Dykes are commonly only a few metres thick and are formed by the intrusion of magma into fissures: some reach the surface to produce an eruption of lava—normally as a fissure eruption.

Eon—The word has two meanings. In this book and in European usage it refers either to all Phanerozoic time or to all Cryptozoic time. In American usage it denotes 1,000,000,000 years (a billion years).

Era—A major division of geological time, including one or more periods: Archaeozoic, Proterozoic, Palaeozoic, Mesozoic, Cainozoic.

Erosion—The wearing away and removal of materials of the earth's crust by natural means.

Eurypterids—Palaeozoic aquatic arthropods, rather like scorpions and probably carnivorous.

Fault—A fracture in the (earth's) crust along which rocks on one side have been displaced relative to the other.

Feldspars—An important group of rock-forming minerals. They are silicates of aluminium with varying amounts of potassium, sodium, calcium and barium.

Foraminifera—A group of single-celled aquatic animals that build a hard case around themselves, either of calcium carbonate or other material.

Fossil—Any evidence of the existence or nature of an organism that lived in ancient times and that has been preserved in materials of the earth's crust by natural means.

Geosyncline—A large trough linear basin in which great thicknesses of sediments may accumulate.

Glacier—A mass of ice, formed by the recrystallisation of snow, that flows forward, or has flowed in the past, under the influence of gravity.

Gneiss—A banded metamorphic rock with alternating layers of unlike minerals: usually equigranular minerals alternate with platy minerals.

Gondwanaland—The grand continent of the southern hemisphere in Permo-Carboniferous time, comprising the assembled continents of South America, Africa-Arabia, India, Australia and Antarctica.

Granite—A coarse-grained, light-coloured igneous rock; it consists chiefly of interlocking grains of quartz, feldspar and mica.

Granodionte—A coarse-grained igneous rock resembling granite but with additional and differing feldspars.

Gutenberg Discontinuity—The surface separating the mantle of the earth from the core below: it lies at a depth of about 2,900 km below the surface.

Guyot—A flat-topped submarine volcano, the summit of which was formerly exposed to marine action and planed away.

Hydrosphere—The water and water vapour present at the surface of the earth—oceans, seas, rivers, lakes, etc.

Igneous Rocks—Rocks formed by the solidification of hot mobile rock material.

Intrusive Rocks—Those igneous rocks that solidify below the earth's surface. They arise from the intrusion of magma into pre-existing rocks where they form magma bodies of a variety of shapes and sizes.

Island Arcs—Chains of islands forming arcs on the surface of the globe. These are the sites of volcanic and earthquake activity and they are usually bordered by a deep oceanic trench on the convex side of the arc (example: Aleutian Islands of Alaska).

Isotope—Some elements are composed of atoms which, while still having identical atomic members and chemical properties, differ in mass.

Laurasia—The grand continent of the northern hemisphere that existed in late Palaeozoic time and comprised North America, northern Europe and Siberia.

Limestone—A sedimentary rock consisting mainly of calcium carbonate.

Lithosphere—The outer shell of the earth lying above the asthenosphere, composed of the crustal rocks.

Magma—Molten rock: generated by the melting of the upper part of the mantle or lower crust. Once formed, magma rises through the crust to give intrusive rocks or is erupted at the surface as lava.

Magnetic Field—The region surrounding a magnet, or magnetic object such as the earth, over which its magnetism has influence.

Mesosphere—The central sphere of the earth's structure, lying below the asthenosphere, it is a relatively rigid mass and contains the inner mantle and the core of the earth.

Metamorphism—The transformation of existing rocks into new types by the action of heat and/or pressure. This process usually takes place at depth in the roots of mountain chains or close to intrusive igneous bodies.

Metazoa—Animals built of many cells, individuals or aggregates of which many have specialised functions.

Meteorite—Mineral or rock material coming to earth from outer space.

Minerals—Naturally occurring crystalline substances with chemical compositions varying between certain fixed limits.

Mobile Belt—An elongate region of the crust suffering volcanic and earthquake activity, subsidence and uplift, with the periodic or ultimate production of uplands or mountains and, in many cases, the intrusion of bathyliths.

Mohorovičić Discontinuity—The surface separating the crust of the earth from the mantle below: it is at a depth of about 70 km below the surface of the continents, and 6-14 km below the floor of the oceans.

Mollusca—The group of animals that includes the clams, snails, chitons and cephalopods.

Nappe—A large body of rocks that has moved a considerable distance over the formations beneath, either by overthrusting or by recumbent folding.

Neptunian Theory—A general theory of the origin of the rocks by deposition from an original ocean. Propounded by A. G. Werner in the eighteenth century.

Orogeny—The process by which the great mountain systems are formed.

Outcrop—Areas where a rock formation occurs at the surface even though it is covered by soil and not exposed.

P-Wave—A compressional earthquake wave.

Pangaea—A. Wegener's name for the super-continent that was formed when the major continental masses were all grouped together, as in Permo-Carboniferous times.

Peneplane—A land surface worn down by erosion to give a flat or gently undulating plain.

Phanerozoic—Referring to Cambrian and later geological time: about 600 million years have passed since the Cambrian period began. Compare Cryptozoic.

Phylum—A major group of plants or animals.

Plate Tectonics—The hypothesis that views the behaviour of the crust of the earth in terms of several rigid plates that are produced by volcanic activity at oceanic ridges and destroyed along the great ocean trenches.

Quartz—A mineral consisting of silicon dioxide, silica. Found in igneous rocks such as granite. Chemically resistant to breakdown, it is a normal constituent of sedimentary rocks such as sandstone.

Regolith—A surface layer of rock fragments formed by the breakdown of the underlying bedrock. It is like soil but may lack the organic material essential to soil.

Regression (Marine)—The gradual retreat of the sea from a land area resulting from progressive uplift of the land or from a change in the actual volume of the sea.

Relative Age—The placing of an event in terms of a time-sequence without regard to age in years. Compare with absolute age.

Sandstone—A cemented or compacted sediment comprising mainly quartz grains.

Schist—A metamorphic rock with strongly developed parallel layering.

Sediment—A loose aggregate of grains deposited from air, water or ice.

Seismology—The study of earthquakes and other earth vibrations.

Shield—The Precambrian crustal mass around which and to some extent on which the younger sedimentary rocks have been deposited; essentially a rather rigid foundation or basement for the continent.

Sial—The continental rocks have a composition rather like that of granite and are formed largely of minerals rich in silica (*SI*al) and alumina (si*AL*). Compare Sima.

Silicate Minerals—Minerals rich in silicon and oxygen, and containing silicon atoms bonded to form atoms of oxygen—the 'silica tetrahedron'.

Sima—Rocks of the crust in the oceanic regions of the earth, composed of minerals rich in silica (*SI*al) and magnesia (si*MA*).

Transform Fault—A large fault set more or less at right angles to an oceanic ridge and separating parts of ocean plates that are growing at different rates.

Transgression (Marine)—Gradual expansion and spreading of a shallow sea resulting in the progressive submergence of land, either by a rise in sea level, or land subsidence.

Trench—Long, narrow troughs forming the deepest parts of the oceans and associated with island arcs.

Tsunami—A large wave in the ocean generated at the time of an earthquake.

Unconformity—The surface separating an overlying younger rock formation from an older underlying formation and representing a period of non-deposition or of erosion.

Uniformitarianism—The principle that the past history of the earth and life is best interpreted in terms of what is known about present natural laws.

Volcanism—The processes leading to and operating during the eruption of volcanic materials from the crust of the earth.

Weathering—The physical breakdown and chemical alteration of rocks by exposure to the atmosphere and/or water, and/or organic matter.

Index